Reading the Comments

Reading the Comments

Likers, Haters, and Manipulators at the Bottom of the Web

Joseph M. Reagle, Jr.

The MIT Press
Cambridge, Massachusetts
London, England

MIT Press books may be purchased at special quantity discounts for business or sales promotional use. For information, please email special_sales@mitpress.mit.edu.

This book was set in Sabon by the MIT Press. Printed and bound in the United States of America.

Library of Congress Cataloging-in-Publication Data

Reagle, Joseph Michael.
Reading the comments : likers, haters, and manipulators at the bottom of the Web / Joseph M. Reagle, Jr.
   pages   cm
Includes bibliographical references and index.
ISBN 978-0-262-02893-6 (hardcover : alk. paper)
1. Online chat groups. 2. Electronic discussion groups. 3. Blogs—Social aspects. 4. Internet—Social aspects. I. Title.
HM1169.R43   2015
302.23′12—dc23
2014034220

10 9 8 7 6 5 4 3 2 1

For my father, Joe Sr.

Contents

Acknowledgments

The sixteenth-century Frenchman Michel de Montaigne wrote that "because there are few who can endure frank criticism without being stung by it, those who venture to criticize us perform a remarkable act of friendship." I thank those who were kind enough to send me constructive criticism—what fan-fiction folks call *concrit*. Whatever merit this book about comments has is dependent on the comments of others. As I tell my students, "Your writing can only be as good as the feedback you get." Of course, it is still the author's responsibility to realize that potential; I hope that I have done so and apologize for where I have fallen short.

I begin my acknowledgments with my colleagues at Northeastern University. Carole Bell, Gregory Goodale, Sarah Jackson, and Brooke Foucault Welles read and commented on one or more chapters. Thank you. Also, Fleura Bardhi gave me some pointers about rating systems, and Alan Zaremba referred me to Montaigne. Other scholars who were generous enough to comment on draft chapters include Thorsten Busch, Eric Goldman, Jeff Loveland, Kelly Page, Trevor Pinch, Andrea Weckerle, and David Weinberger. Devon Powers, a fellow New York University graduate, continues to impress: she read the whole manuscript over a couple of days and sent detailed feedback. The reviewers at the MIT Press were exemplary: they provided timely and productive suggestions to help me write the best book possible. Finding those reviewers is one of the many tasks that MIT Press's Margy Avery undertook on my behalf. Katie Persons reliably responded to my frequent emails. Deborah Cantor-Adams and Rosemary Winfield helped produce and polish the book.

Also, thank you to those who spoke with me about your experiences at the bottom of the Web, three of whom also made suggestions on the manuscript: Brianna Laugher, Foz Meadows, and Denise Paolucci. Four friends also contributed to this project, each in a different way. Melissa Ludtke and I regularly discussed our respective book projects over rosemary french fries. Greg Reagle helped me with the references; it was a delight to collaborate with a fellow geek and a brother. This is Nora Schaddelee's second book with me, and she patiently acted as a sounding board, copy editor, and loving partner. And Casper kept me company through the long days of writing, reminding me when it was time to take a typing break.

Northeastern students are bright and online, and I benefited greatly from discussing this work-in-progress with them in the classroom. Many students were generous enough to do the extra work of sharing their suggestions with me, including Michelle Binus, Stephanie Chaloka, Kitty Cheung, Patrick Christman, Rose DeMaio, Tameka Geaslen, Mark Howitt, Christina LaPenna, Jennifer Lawlor, Adisa Reka, and Sarah Shaker. Anna Glina merits special thanks for reading and giving feedback on my work beyond the syllabus.

I also want to recognize a few institutions that are important to me. I was apprehensive when I first arrived at Northeastern's Department of Communication Studies but it now feels like home, in part because of our chairperson Dale Herbeck and the administrators Angela Chin and Lauren Gillin. I also have had the pleasure of being associated with the Berkman Center for Internet and Society as far back as 1998; it is a wonderful hub for fascinating conversation.

Finally, my thanks to John MacFarlane and the other contributors to the pandoc conversion tool. I write everything in Markdown (books, articles, slides, and email), and this is possible only because of all those who contribute to the text-editing ecosystem.

1

Comment: The Bottom Half of the Web

There's a reason that comments are typically put on the bottom half of the Internet.
—@AvoidComments (Shane Liesegang), Twitter

"Am I ugly?" This question has been asked on YouTube by dozens of young people, and hundreds of thousands of comments, ranging from supportive to insulting, have been left in response. The Web is perfect for this sort of thing, and it has been almost from the start—even if some think it alarming. Over a decade ago, some of the earliest popular exposure that the Web received was through photo-rating sites like HOTorNOT. "Am I Ugly?" videos continue this phenomenon and remain true to YouTube's origins. YouTube was conceived, in part, as a video version of HOTorNOT. YouTube cofounder Jawed Karim was impressed with the site "because it was the first time that someone had designed a Website where anyone could upload content that everyone else could view."[1] But not only could people upload content: others could comment on that sophomoric content. And if the word *sophomoric* seems haughty, Mark Zuckerberg was a Harvard sophomore when he first launched Facemash, his hot-or-not site that used purloined student photos from dormitory directories, what Harvard calls "facebooks."

This uploading and evaluating of content by users is now associated with various theories and buzzwords. (I return to the question of others' physical attractiveness in a later chapter.) *Social media*, like YouTube, are populated by *user-generated content*. Facebook is an example of *Web 2.0* in that it harnesses the power of human networks. The online activity of masses of ordinary people might display the *wisdom of the crowd* or *collective intelligence*. Books on these topics claim that this "changes everything" and is transforming the Internet, markets, freedom, and the world.

Yet I continue to be intrigued by what is happening in the margins—the seemingly modest comment. But what is comment?

As I use the term, *comment* is a genre of communication. In 2010, for instance, YouTube lifted its fifteen-minute limit on videos, and since then, there has been a flurry of ten-hour compilations of geeky, catchy, and annoying audio. Under a hypnotic video of Darth Vader breathing, someone commented: "What am I doing with my life?! 10 hours of breathing!"[2] From this we can see that comment is *communication*, it is social, it is meant to be seen by others, and it is *reactive*: it follows or is in response to something and appears below a post on a blog, a book description on Amazon, or a video on YouTube. Although comment is reactive, it is not always responsive or substantively engaging. Many comments on social news sites are prefaced with the acronym *tl;dr* (too long; didn't read), meaning that the commenter is reacting to a headline or blurb without having read the article. Comment is *short*—often as simple as the click of a button, sometimes measured in characters, but rarely more than a handful of paragraphs. And it is *asynchronous*, meaning that it can be made within seconds, hours, or even days of its provocation. Putting aside future transformations, comment is already present: comment has a long history (some of which I discuss briefly), and it is pervasive. Our world is permeated by comment, and we are the source of its judgment and the object of its scrutiny. There is little novelty in the form of comment itself, but its contemporary ubiquity makes it worthy of careful consideration, especially given online comment's tarnished reputation as something best avoided.

This understanding of comment as communication that is reactive, short, and asynchronous fails to draw a bright line. (I use the term *comment* to speak of the genre and reserve *comments* for an actual plurality of messages.) For instance, at what point does a message become too long to be considered a comment? Unlike a tweet, there is no character limit for a comment, but I focus on communication that is relatively short and can live outside the expectations of real-time interaction. And although these are the rough contours of comment, its essence is best expressed by way of—appropriately enough—an online exhortation: "Don't read the comments." This popular maxim is captured in the tweets of game designer Shane Liesegang. At his account @AvoidComments, he claimed that "there's a reason that comments are typically put on the bottom half of the Internet."[3] There is a lot of dreck down there, but in sifting through the comments, we can learn much about ourselves and the ways that other people seek to exploit the value of our social selves. This book is an exercise in reading (rather than avoiding) comment, and it documents an expedition to the bottom of the Web. I show how comment can inform (via reviews), improve (via feedback), manipulate (via fakes), alienate (via hate), shape (via social comparison), and perplex us. I touch on the historical antecedents of online comment and visit the communities of Amazon reviewers, fan fiction authors, online learners, scammers, free thinkers, and mean kids.

The point of this journey is expressed by an adage often used by media theorist Marshal McLuhan: "We live invested in an electric information environment that is quite as imperceptible to us as water is to fish."[4] Comment is easily seen but invisible to the extent that we take it for granted. Often when comment does make an impression on us, it is a nuisance to be disabled or an offense to be ignored. And even when we do see and appreciate comment, most people have no idea the extent to which it is manipulated. For example, people *like* things on Facebook over four billion times a day, and even these littlest of comments are big business. Scammers are proliferating "*like* farms." When purported pages about cute puppies, brave veterans, and young people stricken with cancer have enough likes, their content is decorated with ads and links to malware sites, or they are sold to others who will do the same.[5] It is easy to appreciate why some recommend that we "never read the comments." Much like California during its gold rush, the bottom half of the Web can be lively and lawless, and it is where many are attempting to make a fortune. Although I do not advocate that everyone read all the comments all the time, I think that it is wise to understand them.

The easiest way to avoid comments is not to have them. Because many sites have disabled their comments, I begin this journey with what gossip teaches about online discussion and why many users and sites are turning away from comment. I argue that disabling comment is a reflection of a platform's growth as users seek intimate serendipity and flee "filtered sludge."

Gossip

The origins of YouTube and Facebook demonstrate that people like to talk about one another: we gossip. Although gossip might seem like a trivial thing, evolutionary psychologist Robin Dunbar argues that it is central to understanding humanity. If you participate in online communities, you might have heard of Dunbar's eponymous number of 150. People invoke *Dunbar's number* when a community (such as an email list whose members used to know nearly everyone else on the list) grows too big. The Web is a big place, and any technology that permits its denizens to communicate with one another has to grapple with the problem of social scale. After the group becomes too large, people complain that the magic has gone. Graffiti and scams proliferate. The known personalities and easy cadence of the group have been replaced by strangers and bickering about unruliness and the need for moderation.

Dunbar did not set out to coin an aphorism about online community. He was seeking to answer the question of why primates, especially humans, are smart—why human brains are about nine times larger, relative to body size, than the brains of other animals. Some have suggested that brain size was related to environment, the use of color vision to find fruit, distances traveled while foraging, or the complex omnivore diet. When Dunbar looked at all of these variables among primates, however, he found no such pattern. But the size of primates' neocortex did correlate with the size of their groups and the time that they spent grooming one another.[6]

A large group is better protected against predation than a small group is, but it also has internal competition for food and mating. Even monkeys can scheme and are sensitive to threats from their peers. Grooming, therefore, is an activity through which alliances are forged and disputes resolved. Experiments with wild vervet monkeys show that they are more likely to pay attention to the distress calls of individuals with whom they recently groomed. But keeping track of who is scratching whose back can be complicated. In a group of twenty, there are nineteen direct relationships and 171 third-party relationships, so as group size increases, so does the complexity of the network and the time that primates spend grooming one another. According to Dunbar, primates' large brains are the result of an evolutionary race of alliances through social grooming.

For humans, social grooming includes language. Because larger groups require more efficient means of forging alliances, gossip circulates information about others in the social networks in which they exist. While the practice of talking about others (including rumors and bathroom graffiti, or latrinalia) is more interesting and complex than I can address here, I understand gossip simply as "evaluative social chat."[7] And the alliances that result from sharing opinions about others can be Machiavellian. On a television reality show, for example, Sandy might realize that John's seeming betrayal of Alice could itself be a lie. Dunbar argues that gossip requires a sophisticated type of social cognition known as the *theory of mind* through which we infer the mental states of others. Even four-year-old children demonstrate second-order intentionality: the child has a belief about what someone else wants. Adults can negotiate fourth- or even fifth-order intentionality. This is amusingly demonstrated in a scene in *The Princess Bride* where two opponents engage in a battle of wits. The "Man in Black" poisons one of two goblets of wine, and Vizzini must choose and drink from the safe one. Vizzini begins his chain of deduction with the assumption that "A clever man would put the poison into his own goblet, because he would know that only a great fool would reach for what he was given. I am not a great fool, so I can clearly not choose the wine in front of you. But you must have known I was not a great fool. You would have counted on it, so I can clearly not choose the wine in front of me." This type of inference requires a "clever man" with a big brain.

In any case, Dunbar's number of 150 is, roughly, the cognitive limit of how many relationships humans can maintain given their complexity (such as "the enemy of my enemy is my friend"). However, Dunbar proposed multiple tiers, from the family up to the tribe of a couple thousand people. The rough size of the *clan*—the individuals that a person keeps in contact with and can track relationships for—is 150 people. This is roughly the size of early farming communities, modern planters in Indonesia and the Philippines, and contemporary Hutterites. It is roughly what the Church of England concluded to be the ideal size for a congregation and the number of soldiers in a military company. Also, using the birth rates observed in hunter-gatherer or peasant societies, this number corresponds to about five generations, which is as far back in time as anyone living can remember: "only within the circle of individuals defined by those relationships can you specify who is whose cousin, and who is merely an acquaintance."[8]

How is Dunbar's number related to comment? It provides an unexpected clue to why comment frequently fails on the Web.

Disabling Comments

In the blogging domain, there is little that Dave Winer has not written code for, started a company around, or opined about. He often is credited with deploying the first blog comments in 1998. (Another contender for this claim is Bruce Ableson at Open Diary.[9]) Despite his quick smile and easy manner, Winer ends up in a lot of online arguments. In 2001, he was described in the *New York Times* as someone who is "not shy about ruffling the big names in high technology."[10] Winer is opinionated and passionate. He also is willing to say that something "sucks" or to call someone an idiot. As I will discuss further, *drama genres* of comment, such as sites where people can ask others questions or make lists of things to avoid, lend themselves to conflict. Certain personalities also seemingly attract conflict. In 2006, when Winer announced that he would stop blogging before the end of the year, a group of antagonists created a countdown clock that they said would continue "until he shuts up." (He continued blogging.) Critics expressed their antagonism in comments to his blog, and he repeatedly considered disabling the comments altogether. In 2007, he wrote that he did not think comments were an essential part of a blog, especially if they "interfere with the natural expression of the unedited voice of an individual":

We already had mail lists before we had blogs. The whole notion that blogs should evolve to become mail lists seems to waste the blogs. Comments are very much mail-list-like things. A few voices can drown out all others. The cool thing about blogs is that while they may be quiet, and it may be hard to find what you're looking for, at least you can say what you think without being shouted down. This makes it possible for unpopular ideas to be expressed.[11]

In 2010, Winer developed this idea further, arguing that blog comments should be short and about the blog posting (or *responsive*, using my term). They ought not be digressive or overly long. He proposed a system that allows comments of less than a thousand characters to be invisibly submitted within twenty-four hours: "After the commenting period is over, the comments would become visible, and no further comments would be permitted." Those who wished to respond later or at greater length could do so on their own blogs. This idea was supported by the *trackback* feature of many blogs. If I respond to a blog post by Winer with a post on my own blog, for example, then my blogging platform would inform Winer's blog service of my response. Winer's blog entry would then include a link to my own. Winer's blog "tracks back" to those who respond. Trackbacks were seen as a way to complement or replace comments, but they have largely fallen into disuse after their abuse by spammers. For Winer, neither blog comments nor tweets are appropriate for conversation. In 2012, Winer, the person who often is credited with first enabling blog comments, disabled them from his own blog—seemingly forever.[12]

As Mathew Ingram, a technology writer, noted about a 2007 fracas over Winer's "Why Facebook Sucks" posting, Winer's approach sometimes "brings the hate."[13] But Winer's experience is not unusual. After the halcyon days of blogging, many bloggers abandoned their sites or shuttered their comments. Some popular sites (including *Boing Boing* in 2003, the *Washington Post* in 2006, *Engadget* in 2010, and *Popular Science* in 2013) have turned off their comments for extended periods. In 2013, Rob Beschizza, managing editor of *Boing Boing*, tweeted that he may do so again "for good," perhaps after inappropriate comments were made on a posting about the death of a friend.[14] *Boing Boing* began as a paper zine in the late 1980s and went online in 1995. In its early days, it was like an informational swap meet among friends, but by the new millennium, it had become too popular to serve as an unfettered venue for sharing and gossiping among friends and has struggled with this fact ever since.

However, there are two other responses to unruly comments beyond disabling or ignoring them. Website managers can attempt to *fortify* their commenting system and Website users can *relocate* in search of what I call *intimate serendipity*.

Fortifying Comment Systems

By *fortify*, I mean to make the system more resistant to abuse. Some sites require users to perform a task (like typing in text that has been distorted) before leaving a comment so as to minimize abuse. However, abusers often match the cleverness of the challenge or farm out the task to low-cost workers on the other side of the world. Many sites permit readers

HOW THE NET HAS CHANGED OUR LIFE

ISN'T IT GREAT THAT NOWADAYS EVERYBODY HAS A CHANCE TO BE HEARD?

to filter comments based on ratings from other users who act as moderators, such as at the nerdy news site Slashdot. This site also uses metamoderation, whereby others' moderations can be rated as fair or unfair. Even so, people sometimes complain that a *cabal* of moderators has taken over and is abusing the system. That is, a group of users collude to promote one another's postings and standing. One can often find someone complaining in a story's comments that their submission and summary of the story was better and earlier but that it was ignored because she was not part of a clique.

Facebook and Google+ have required users to use their real names. While Facebook has been relatively lax in enforcement, Google+ was quite strict at its start but stepped back from the requirement in 2014. (My dog has a Facebook page, but it is under his real name.) Such social networks are then able to leverage their identity policies and reach by providing authentication and commenting services for others. *Slate* adopted Facebook's 2011 "Comments Box" service, and Farhad Manjoo, a staff writer at *Slate*, was pleased that Facebook knew real names and that comments could be seen on Facebook by the commenter's friends and family: "This introduces to the Web one of the most important offline rules for etiquette: Don't say anything that you'd be ashamed to say in front of your mom."[15] MG Siegler, a blogger at *TechCrunch*, noticed that since they "flipped the switch" and adopted Facebook's service there had been a large drop in comments, but "this is completely expected and definitely not a bad thing." Before, a post might get hundreds of comments, half of which were "weak to poor" and half of those "pure trollish nonsense." With the new system, a similar post might receive about a hundred comments, half of which "are actually coherent thoughts in response to the post itself—you know, what a comment is supposed to be."[16] Others claimed that real-name policies suppressed anonymous speech and were incompatible with the multiple identities that we maintain in life. Furthermore, using centralized commenting systems cedes ever more autonomy and privacy to the likes of Google, Facebook, and other comment-specific services like Disqus (used by 750,000 sites, including CNN's Website) and Livefyre (used by the BBC and the *New York Times*).

Sometimes, relatively simple approaches do the trick. The link-sharing and discussion site MetaFilter requires a one-time $5 membership fee.

This fee, its strong community norms, and occasional moderation for flagrant abuse seem to have fostered a robust and civil culture. Newspapers, too, have experimented with asking readers to subscribe or pay a small fee to comment. In keeping with Dunbar's insight, Clay Shirky, author and NYU professor, is fond of examples of communities whose size is purposefully limited. Ten years ago, he wrote about an email list that removed the oldest subscribed member to make room for the newest one. Another list's periodic purge was inspired by a supposed neighborhood hot tub that was accessed by a key-coded gate lock: people could give the code to friends, but when the bathers became too rowdy or the tub was overcrowded, the owner simply changed the code and gave the new one to his immediate friends under the same policy.

A decade later, Shirky continued to stress that "intimacy doesn't scale." In a 2013 talk entitled "Why Do Comments Suck?," he put it simply: "Comment systems can be good, big, cheap—pick two."[17] Many sites with comments seek a large audience. They "want lots of people to forward the article to a million friends, shut up and then read another article." Sites that treat their users as community members (through smaller size or careful moderation) tend to have better comments. This is a good insight but easier said than done: good communities tend to grow. This is the paradox of their success. People then often relocate to another site without a good understanding of what went wrong (except that it went "downhill") or what they were looking for in the first place.

Twitter and the Search for Intimate Serendipity

We now have a toolbox of tactics for resisting comment abuse, and they often are good enough—for a time. But some communities struggle as they experiment with finding a system that is appropriate to their changing size and circumstances. Those not satisfied with the changes often leave and *relocate* to a new media platform in search of what I call *intimate serendipity*. When I went to blogging get-togethers in 2003, it was with a dozen of like-minded enthusiasts: I met interesting people and we had good conversations. Over a decade later, going to a meeting for people who post comments to the Web seems passé. (Today almost any gathering could qualify as such a meeting.) After a network of people (online or otherwise) becomes popular, people want to bring their friends.

At first, this is great. The value of a network increases significantly with each new node. A network of five phones permits ten connections; doubling the phones to ten permits forty-five possible connections. As Dunbar notes, however, at some point the scale of networks overwhelms the participants. First, we ask, "Who brought that guy to the party?" Second, the network becomes a target for those who wish to exploit it via spam and manipulation.

I sometimes ask my students if the parade of Web platforms is over (i.e., Geocities, Myspace, Facebook, Twitter, and Google+). Given the difficulties involved in leaving a service (because content and connections must be abandoned) and the profusion of niche networks, some might think that there is little need for anything new. But people do relocate when an existing platform becomes overly populated by jerks, spammers, and ads or overly constrained by controls and filters. People often want a network where intimacy and serendipity are possible. Although there are sites where being anonymous and a jerk is the norm, many people want to express their authentic selves without fearing attacks, manipulation, or unusual exposure and while remaining open to things that surprise and delight. When they can't do so, they move, as seen in the migrations of geeks between social news sites like Slashdot, Digg, Reddit, and Hacker News.

By 2007, three years after its founding, Digg was being criticized as a system that was rife with manipulation by supposed "bury brigades" that suppressed others' stories. At the same time, the company was trying to become financially viable or even profitable. As users began to leave, the site exacerbated discontent with the introduction of unpopular changes, including new comment systems and site designs. By 2010, the site was on its deathbed; while the service limped on, and the name had been affixed to various "reboots," its founder, staff, and community were gone. Many who left Digg relocated to Reddit. Correspondingly, those who had been at Reddit, especially the technically inclined, lamented that the site had become too big and diluted. When Paul Graham launched Hacker News in 2007, the intention was to replicate the early Reddit days, and some early contributors to Reddit followed.

The irony is that success sometimes, seemingly, brings about a comment system's demise. In 2009, law professor and civic reformer Lawrence Lessig announced that he was retiring his blog. Lessig's problem

was not with haters but with the love. He is an influential legal scholar, a successful author, and a founder of significant cultural, academic, and civic organizations. He has argued against copyright extensions before the U.S. Supreme Court, is a founder of Creative Commons, and popularized the notion of "free culture" with a successful book. However, a growing family, a shift in research focus, and the "increasingly technical burden to maintaining a blog" were too much. He and his volunteers tried to keep up, but a third of the thirty thousand comments on his blog were likely "fraudsters," and online casino spam was growing. Google stopped indexing the site at one point. However, he was "still trying to understand twitter."[18] In fact, lots of people were moving to Twitter. For a time, it gave its users intimate serendipity.

At its beginning in 2006, Twitter felt edgy and intimate. It was not uncommon to find users flush with an encounter with a (minor or major) celebrity.[19] Also, people (especially the famous) were thrilled to be able to authentically express themselves. People like talking about themselves, and Twitter appeared to be a safe space to do so. Research indicates that people spend 30 to 40 percent of their interactions telling others about their subjective experiences, so it is not surprising that researchers found that 41 percent of the tweets in their study were of the "me now" variety.[20] Two Harvard University neuroscientists concluded that "disclosing information about the self is intrinsically rewarding" because it triggers regions of the brain that are associated with the mesolimbic dopamine reward system. In experiments in which subjects could choose to speak about themselves or factual matters, people chose to speak about themselves the majority of the time. When these choices were associated with small payments, people were willing to pay an average 17 percent "tax" to talk about themselves: "Just as monkeys are willing to forgo juice rewards to view dominant groupmates and college students are willing to give up money to view attractive members of the opposite sex, our participants were willing to forgo money to think and talk about themselves."[21]

Trent Reznor, the digitally progressive artist behind the band Nine Inch Nails, initially took advantage of this self-disclosure. At first, Twitter allowed him to "lower the curtain a bit and let you see more of my personality," more so than what he could do on the forums at his site nin.com:

The problem with really getting engaged in a community is getting through the clutter and noise. In a closed environment like nin.com a lot of this can be moderated away, or code can be implemented to make it more difficult for troublemakers to persist. It's tedious and feels like wasted energy doing that shit, but some people exist to ruin it for others—and they are the ones who have nothing better to do with their time. Example: on nin.com, there's 3–4 different people that each send me between 50–100 message per day of delusional, often threatening nonsense. We can delete them, but they just sign back up and start again. Yes, we are implementing several changes to address this, but the point is it quickly gets very old weeding through that stuff.[22]

At Twitter, Reznor "approached that as a place to be less formal and more off-the-cuff, honest and 'human.'" Some of his tweets were about a new love in his life, which some fans found incompatible with his earlier dark and alienated music: "I'm not the same person I was in 1994 (and I'm happy about that)." This worsened in 2009, when the obsessively vitriolic fans he called "the Metal Sludge contingency" discovered Twitter; he vowed to quit and has tweeted only sporadically since. Unlike email, where addresses can be kept private and messages can be filtered, a Twitter handle is public. Such openness worked as long as the community was sufficiently small, but by 2009, the tweet bomb had arrived: people were spamming others. Twitter now permits users to block or label others as spammers, and serious users use third-party apps with powerful filters. But new Twitter accounts are easily created or cheaply purchased by the thousands, as I discuss in a later chapter. By summer 2013, people were using a "block bot" to filter out "general bigots, assholes and fools."[23] Yet, a centralized block list is controversial for many reasons, including who decides who goes on the list.

Many social platforms move from intimate serendipity toward filtered sludge, and some manage it better than others. As investors begin to demand a return on their investment, the sites themselves are tempted to alienate their users with ever more intrusive filtering and ads. This is the life cycle of a social media platform. Although I expected Twitter to face the same dynamic, I was surprised that its crisis began with a case of humanitarian advocacy.

The Tweet Bomb and the Loss of Intimate Serendipity

In spring 2012, I was teaching an introductory media course that covered various models of media persuasion, including the *opinion-leader* model. In this view, influence flows from the media to opinion leaders and then from opinion leaders to other people. Fortunately, there was an example on hand that many of my students knew about: the campaign to have Joseph Kony arrested for war crimes in Africa, including his use of child soldiers. This campaign was led by the organization Invisible Children, which targeted its message to young people in the West. In 2012, the group launched a social media campaign to have Kony captured by the year's end. In addition to creating the online film *Kony 2012*, campaigners tweet-bombed celebrities; Rihanna, Justin Bieber, and Oprah retweeted for the cause. As media scholar and Internet activist Ethan Zuckerman notes, however, this "attention philanthropy" does not lend itself to indefinite repetition: "Oprah has a great deal of a valuable commodity—attention—and the incremental cost of her spending that attention to call attention to a cause is minimal.... In the long run, if she tweets about every campaign her fans want her to promote, she'll likely start to lose her audience—the incremental cost may be small, but the cumulative cost could be very high."[24]

A month after the Kony campaign, Bachir ("Athene") Boumaaza (an Internet celebrity known for his gaming prowess, YouTube channel, and social activism) attempted the same tactic as part of Operation Share-Craft. Athene and his collaborator Reese Leysen were attempting to leverage Athene's significant online following to ameliorate hunger in the Horn of Africa. Much like the Invisible Children campaign, Operation ShareCraft sought the attention of prominent social media personalities, including Xeni Jardin, a founding contributor of the popular blog *Boing Boing*. However, Jardin's attention was elsewhere. Four months earlier, she had scheduled an appointment for a mammogram and had decided that sharing her experience would allay her anxiety and perhaps encourage others:

I would tweet this new thing, like I do with lots of new things, and make the unknown and new feel less so. Maybe by doing so ... other women like me who'd never done this would also feel like it was less weird, less scary, more normal and worth doing without hesitation. I'd crack some 140-character jokes. I'd make fun of myself and others. I would Instagram my mammogram.[25]

Apparently, this is not unusual. In his book about "social communication in the Twitter Age," Dhiraj Murthy dedicates a chapter to health, focusing on cancer. He writes that although many cancer-related tweets are either about charities or news of treatments, ordinary people share their own experiences, often at the intersection of the banal and the profound. People who were anxiously sitting in a doctor's waiting room were unlikely to write a blog entry about it but would "pick up their smartphone for 45 seconds and Tweet about it."[26]

After Xeni Jardin received the results of her mammogram, she tweeted "I have breast cancer. I am in good hands. There is a long road ahead and it leads to happiness and a cancer-free, long, healthy life."[27] Many people found the reports of her ongoing treatment compelling, and fellow patients and survivors used Twitter to exchange information and support. Although Operation ShareCraft's tweet-bomb plan was to ask celebrities "to spread the word or support the cause in any way possible," Jardin found the group's mass entreaties to be unseemly.[28] She tweeted to Athene and Reese that this was "totally tone-deaf and inappropriate" and complained of "Getting tons of SRY U HAZ CANSUR PLZ DON'T DIE XENI PS HELP US END HUNGER IN TEH HORN OF AFRICA! KTHXBYE spam tweets. Fuck all of you, srsly."[29] A slew of hateful and misogynist tweets followed. Athene and others apologized, explaining that "trolls tried to sabotage the event," yet Jardin continued to receive messages to the effect that "All these people sacrifice their free time to raise awarness and do good. If you dont wanna get tweet bombed dont be on Twitter."[30] Jardin concluded that she had no ill will toward the charity and that this had been the "Strangest griefer/troll storm I've ever experienced here. Hope it's not indicator of Twitter's future."

Comment in Context

This book isn't about the future of Twitter, blogs, or YouTube, but it is about comment in the age of the Web. (Nor is this book about looking for comments from time-travelers from the future, as some researchers have done![31]) Similarly, it's not about how social media have transformed politics, journalism, or global relations. This book is about the stuff in the margins—the things that ordinary people encounter in daily life. YouTube's Jawed Karim recognized the importance of user-generated content at HOTorNOT, but many continue to overlook the comments on such content. Indeed, the unsavory aspects of online comment have prompted many to turn a blind eye to the "bottom half of the Internet" and to advise visitors, "Don't read the comments." And although headlines publicize the IPOs and purchases of related Websites, many are unaware of the illicit markets in which followers, reviewers, and commenters are bought and sold.

In addition to defining comment as reactive, short, and asynchronous, it also is useful to consider its context. A comment is about something— an *object* or a topic, such as a book. A comment has a *source* or author, who might be identified or anonymous. It has an *audience*—the people who are the intended readers of the comment. The *content* of a comment might be prose, a verbal aside, or a rating. Even clicking a *+1* or a *like* button is a comment. Finally, the *intentions* and *effects* of comments are important. A comment can affect another's status (for example, a diploma is a comment about academic standing, and gossip is a comment about social standing), it can help others make decisions (such as "The food here is great"), or it can alter a person's behavior (for example, by providing feedback about someone's actions). Much drama can ensue when the context of comments is ambiguous or transgressed, such as when a note to a friend at school is intercepted and becomes known to everyone in the classroom.

Two chapters in this book focus on the intentions and effects of comment: informing (such as with a review that helps readers choose a product or service) and providing feedback (such as with remarks that help people improve their lives). With respect to reviews, in chapter 2, I note that comment today inherits many reviewing modes from the past. The Web contains the legacies of early twentieth-century engineers who were preoccupied with comparative analysis and of two brothers who sought to sell more tires by assigning a number of stars to hotels and restaurants. I also discuss the *likers*, who share recommendations that are rooted in love and experience; the crowd, which shares its particular, peculiar, kind of wisdom; and the critics, who highlight and connect with analysis and insight. All forms of writing that have gone before are present on the Web—and at a very large scale. These types of comment existed before the twenty-first century, but never were they available in such great numbers or were they as easily accessible as they are today.

Because comments have become one of the most valued commodities of the day, they are subject to much manipulation. In chapter 3, I review the research on fake reviews and discuss some prominent cases of fakery. I distinguish between makers, fakers, and takers and discuss the dynamics and techniques of online fakery. I argue that relying on social networks (consuming the recommendations of our friends rather than strangers) will not solve the problem of manipulation but will tempt us to become manipulators ourselves.

In chapter 4, I focus on feedback, which is a practice that has great potential to go awry. This is especially so in the case of what I call *tweak critique*, such as altering someone else's photo to improve its composition. Because it is now easy to solicit and give feedback, quick and unthinking comment can easily bruise and can blur the distinctions between criticism, feedback, and review, which results in contention and controversy. Similarly, the scope and scale of feedback have changed: feedback to one person can be seen by many and can be unsolicited and unwanted. Feedback requires the giver to be careful and the receiver to perform what scholars call "emotion work."

Unfortunately, online conflicts are not always between well-intended commenters who try to be civil. Although comment is a type of communication that permits us to be helpful, friendly, and encouraging to others, it can lead to feelings of frustration and alienation. In chapter 5, I describe the trolls and haters, bully battles, and misogyny that often are encountered online, and I frame this culture with what is known about the effects of anonymity, deindividuation, and depersonalization. Alienating and hateful comment is not likely to go away anytime soon, but we are beginning to give greater attention to how to respond to it at technical and communal levels.

Online comment is reactive and short, and these characteristics affect writers and readers in a couple of ways. First, our reaction to things (be it a comment, an answer to a question, or the liking of a photo) has come to be seen as a way in which we define ourselves. And the ways that others react to those reactions (such as by retweeting them) are seen as a valuation of those selves. Second, comment's shortness and ubiquity mean that attention is easily and often drawn online. And the fact that these comments can be counted and tracked also affects how people value themselves and others. Chapter 6 asks how this is shaping us. How does the nonstop stream of our own and others' photographs and status updates affect self-esteem and well-being? Are the short and asynchronous bursts of comment that are processed throughout the day affecting our ability to concentrate? Have the quantification and ranking of social relations gone too far? These are complex questions without easy answers.

Before concluding the book, I show in chapter 7 how puzzling comment is. In addition to informing, improving, alienating, manipulating, and shaping, these short and asynchronous messages can bemuse: they can be slapdash, confusing, amusing, revealing, and weird. Because comment is reactive, it is inherently contextual. Yet it also is *hypotextual* (that is, undertextual), shedding context with ease and prompting the comment of *WTF?!?* (what the fuck?). From this confusion and weirdness, however, we can learn about the advantages of moving first, the intricacies of communication, the science of rating systems, and the challenges of not losing context at the bottom of the Web.

2

Informed: "I Don't Know. I Gotta Get the Best One"

Being a consumer is like a job. You have to make sure you get the best one. If you get a Blu-ray player, you gotta do research.... You gotta go on Amazon and read a really long review written by an insane person who's been dead for months ... because he shot his wife and then himself after explaining to you that the remote is counter-intuitive. "It's got really small buttons on the remote," he said ... before he murder-suicided his whole family. And now you're reading it and going, "I don't know. I don't know which one to get, I don't know. I gotta get the best one." Who are you, the King of Siam, that you should get the best one ever? Who cares? They're all the same, these machines. They're all made from the same Asian suffering. There's no difference.

—Louis C.K., "Late Show: Part 1," *Louie*, season 3, episode 10, August 30, 2012

Despite the name, Boston's Micro Center isn't small—nor is it in Boston. It is, instead, an electronics supermarket on the Cambridge shore of the Charles River. It attracts customers from all over eastern Massachusetts, but I live a few minutes away. Once, while browsing the aisles, I came across an inexpensive accessory for my camera. I wasn't happy with the included strap and for a mere $6 this one would give me added security against dropping it. But was it any good? I left my phone at home, and so in a quiet corner I found an online laptop and quickly pulled up Amazon. The strap could be had for a few dollars less and, more importantly, it had a few decent reviews: it's not a lemon. Good, except that a sales associate looked over my shoulder and asked if I needed any help. Embarrassed, I explained that I will indeed be buying the product and was only checking the reviews.

Even when I plan to purchase something in a store, I feel uneasy about buying blind. With Amazon's "Price Check" app on my phone, a quick photo of a bar code reveals all. As a reviewer of the application noted,

"I use this app not for price (unless a huge difference), but to check the product review. It's so handy! I can avoid buying things [the] amazon community doesn't give a stamp of approval." In fact, a third of surveyed adult cell phone owners said that they used their devices to look up reviews and prices while in a store during the 2011 holiday shopping season.[1] Such people likely include *maximizers,* a term popularized by Barry Schwartz in his book *The Paradox of Choice.* Whereas a *satisficer* can settle for good enough, maximizers must be assured that every decision is optimal.[2] They spend hours reading reviews and feel disappointed when an item falls short of expectations or is surpassed by a new model. They suffer from the fear that they could have made a better decision; this is the paradox of increased information and choice.

Today maximizers have an extraordinary amount of information available to them. Sites like Yelp and Amazon offer ratings in the form of stars. These ratings can be accompanied by reviews that are often haphazard but sometimes astoundingly detailed. Reviews can be professional, such as those in *Consumer Reports,* or amateur, such as those at Yelp, Amazon, and everywhere else on the Web. Because confidence correlates with large numbers, some sites distill the ratings, such as Rotten Tomatoes' "freshness" percentage, Yelp's average rating, and Amazon's histogram. A *bimodal distribution* in which most ratings are either zero or five stars is a sign of controversy. Reviews themselves can be reviewed as "helpful" and commented upon. Forums and lists provide additional ways for people to discuss and perhaps even form a community. Unsure of the quality of tangible goods, reviewers can view photos and videos of products. On YouTube, these reviews serve as a way for reviewers to fashion their

identity as a helpful expert (for example, with comments such as "This is my favorite mascara and here are my application tips") or conspicuous consumer ("Let's drop test the latest gadget").

This information is varied and rich but it is not novel simply because it appears on the screen. An essential function of comment is to inform: we express our thoughts for the benefit of others, and others seek them out to understand and make decisions. As noted in chapter 1, we have been gossiping about each other for a long time, and we can even understand gossip as a part of what makes us human. Similarly, the desire to comment on a written text is as old as writing. As soon as humans began writing, our scribbles have been a source of confusion and contention and have necessitated commentary. Perhaps the earliest instance of this is the Babylonians' dictionaries of Sumerian. Yet we have needed help with more than just foreign languages. Because early writing lacked many of the conveniences that we take for granted today—such as vowels, punctuation, and spaces between words—the ancients needed help in deciphering their cherished texts. Hence, they developed conventions for annotating these works and these (also ancient) annotations are known as *scholia*.[3]

Tom Standage's 2013 book *Writing on the Wall: Social Media—the First Two Thousand Years* is a delightful corrective for our myopic tendency to think that the new is necessarily novel. For instance, the Romans wrote on their friends' walls (literally), and Martin Luther "went viral" thanks to the extraordinary information technology of the day: the printing press. Pope Leo X likened the spread of Luther's protests to a "plague and cancerous disease."[4] When considering reviews (comments that inform people), it is also worthwhile to consider the past. So I begin this journey to the bottom of the Web with a brief historical excursion. Many books have been written about criticism and review, but I suffice with a brief discussion of what is most often seen online, touching on the origins of the review, gold stars, likers, the crowd, and the critics now found on the Web. This discussion is from the perspective of a gadget addict who also is addicted to gadget reviews. For example, as a *maximizer*, I spend far too much time in search of the perfect product and often guiltily recall comedian Louis C.K.'s question: "Who are you, the King of Siam, that you should get the best one ever?" Reviews have been around for a while now, but never before were they so accessible that we'd regret our purchase the next day upon reading someone else's comment.

The Review

When buying a camera, the new owner is advised to read the manual and, if overwhelmed and confused by the manual, to purchase another guide (sometimes called a "missing manual") with more detailed instructions about how to use the gadget's many functions. That is, an expert is needed to help the user to decipher and apply the product's manual. Photographer Gary L. Friedman, for example, sells ebooks with "professional insights." These texts are a hybrid of a professional review and user guide. Content seemingly calls forth even more content, which is a recurrent theme in the history of media.

During the Enlightenment, cloistered scholars slaving over the annotation of ancient authorities gave way to the likes of John Locke, Voltaire, and Isaac Newton; new thoughts and works abounded. However, the glut of work from lesser thinkers led Gottfried Leibniz, Newton's contemporary and fellow inventor of calculus, to complain of the "horrible mass of books that keeps on growing." He feared that there would be no end "to books continuing to increase in number," and he was right.[5] By the eighteenth century, print's proliferation called into being new forms of commentary—ironically, more print—that we now take for granted. The seventeenth century's comprehensive indexes and collections of abstracts were followed by more discriminating reviews in the eighteenth century. The French *Encyclopédie* and other reference works are illustrative of a new reading public and their desire to have the mass of knowledge made sensible and accessible. Although some might not view reference works as commentary, early reference work compilers were an opinionated bunch. In Dennis de Coetlogon's *A Universal History of Arts and Sciences* (1745), the "Geography" article begins with the nation of France because it is "the first in rank" and "the most fertile, the most agreeable, and the most powerful in Europe."[6]

Print's proliferation accompanied the emergence of a new class of wealthy literates, including merchants and bankers. As chronicled by the German philosopher Jürgen Habermas, the new bourgeois, this "reading public," constituted a "public sphere" in which all topics were discussed without deference to the authority of the ancients or of contemporary rulers. In fact, this led Charles II of England to issue a "Proclamation for the Suppression of the Coffee-Houses" in 1675.[7] (Today we have check-in apps such as Foursquare that permit users to proclaim their fondness for a café in hopes of becoming its "mayor.") Despite the proclamation, caffeinated commentary was not easily quieted. In the final months of the reign of Louis XVI of France, his government solicited lists of grievances (*Cahiers de Doléances*) from his subjects as suggestions for reform. This backfired on Louis because, when written down, the complaints hastened his end and the beginning of the French Revolution.

The new literates' tastes were not limited to the civic and natural domains. As Frank Donoghue argues in *The Fame Machine: Book Reviewing and Eighteenth-Century Literary Careers*, all types of authors (now severed from the apron strings of patronage) sought to make careers in a marketplace that was characterized by an uneasy mix of contention and cooperation between authors, reviewers, and readers. London's *Monthly Review*, established in 1749, initially conceived of itself as being "serviceable for such as would choose to have some idea of the book before they lay out their money or time on it."[8] Many competitors soon followed, especially *The Critical Review*, and the rivalry between these two reviews shaped the emergence of this genre as much as anything else.

By the beginning of the twentieth century, the "penny dreadful" in the United Kingdom and "dime novel" in the United States exemplified a further increase in popular literacy and a burgeoning consumer culture. Over five thousand new books and editions were published in 1896, the year in which the first "New York Times Book Review" was released (initially as the "Saturday Book Review Supplement"). At some point, the genre of the book review became so popular that it became a primary school assignment, and as early as 1885, teachers' assignments included asking each student to "prepare a summary of some chapter" from their reading. In 1919, Wid Gunning, a twenty-nine-year-old film fan, began publishing *Wid's Films and Film Folk* with reviews of over fifty silent pictures. In 1925, *The New Yorker* published its first issue, setting the mold for modern commentary and criticism.[9]

One lesson that can be drawn from this brief history of reviews is that *comment leads to more comment.* This is still true today, and the pace of change online can be understood as one generation of information management being overwhelmed and complemented by the latest. In the late 1990s, Web "portals" were the rage, and companies raced to dominate the collection and organize information for the users. In the new millennium, hierarchical directories have been supplanted by tags, keywords affixed to a piece of digital content by the users. As David Weinberger argues in *Everything Is Miscellaneous*, tags are the embodiment of a new "digital disorder" in which the organization of information can be fluid, ad hoc, and disposable.[10] People try to stay abreast of it all by consuming reports of trending tags and presentations of "tag clouds" in which terms are rendered in relative proportion to their popularity. Tags have even become a battleground where different factions compete to coopt the tags of their opponents. Online feminists, for instance, post ironic tweets under #INeedMasculismBecause, and masculinists do the same under #tellafeministThankYou. (I return to *hash crashing* in chapter 5.) In short, comment begets comment.

The Stars

In a four-star review of "Bacon Flavored Toothpaste," an Amazon reviewer wrote, "The stuff tastes horrible, but that was why I bought it as a gag gift haha. I gave four stars because besides bacon and mint, another taste was present like a manufacturing ... chemical ... I don't know ... bad taste that didn't belong."[11] That is, the reviewer liked the expected surprise taste of bacon but subtracted a star for the unexpected taste of the unknown. We are fortunate that Amazon permits its commenters to explain their allotment of five stars, but in the past few years we have seemingly undergone an even greater compression of expression. If a comment is stripped away of all that is superfluous, a kernel of disposition remains. Do you *like* this thing? Do you agree (+1)? Do you wish to share it? This concept underlies Facebook's *like* button. On its release in April 2010, Mark Zuckerberg predicted that this newest form of (attenuated) commentary would be used over one billion times within its first twenty-four hours. Facebook never substantiated its original boast, but it noted that within the first week fifty thousand Websites implemented its new social plugins.[12] Hundreds of millions of people at least saw the new button. Yet a better measure of its impact may be how quickly its competitors moved to provide alternatives. In August 2011, Google launched the *+1* button for its social network. By 2012, Web pages were riddled with social buttons.

In January 2013, the synth-pop group The Knife released a nine-minute music video for the lead track from their forthcoming album. Beneath the video was a button for sharing the content on social media. Clicking the "Add This" button led users to top-level buttons for Facebook, LinkedIn, Twitter, email, and "More." Hitting "More" revealed buttons for print, Gmail, StumbleUpon, Favorites, Blogger, Tumbler, Pinterest, and "More." The final "More" provided buttons for over 337 different sites! The button glut prompted a prominent designer to advocate that people "Sweep the Sleaze" from their Webpages.[13]

Of course, stars and *like* buttons are not innovations of the twenty-first century. School children have long received gold stars for good spelling and handwriting. Some foodies might know that Michelin rates restaurants with stars without knowing they make tires, but it has done both since 1900. The earliest significant guides appeared at the end of the nineteenth century—the *Hachette* in France and *Baedeker* in Germany—and were designed for the railroad traveler, with suggestions for carriage excursions. But the car and its tires were the engine of the restaurant review and the source of the now ubiquitous stars.

In his book *The Michelin Men: Driving an Empire*, Herbert Lottman recounts the history of the company that was started by the Michelin brothers, including the synergy of the new automobile with increased literacy and leisure. In 1900, Michelin printed its first guide, and stars were used only to indicate the class or cost of hotel accommodation, a convention that was used by existing train guides. The guide also included symbols for parking garages, type of motor oil, repair shops, and darkrooms. The following year it indicated whether hotels had flush toilets, bathtubs, and showers. However, it wasn't until 1925 that Michelin introduced the three stars that we know today, though their eventual significance was not yet appreciated. Between 1925 and 1930, different systems were used in different guides: Michelin distinguished between stand-alone and hotel restaurants, between Paris and the provinces, and used a five-star system in some of its guides. In the 1930s, the three-star system began its ascendancy with a charmingly modest meaning: a restaurant or hotel could be "Worth a detour" (two stars) or "Worth a journey" (three stars). Today, a single star denotes "A good table in its community," two stars an "Excellent table: worth a detour," and three stars "One of the best tables in France; worth the trip."[14]

Confusion over the number and meaning of stars is a topic that I return to in chapter 7, but two other aspects of the history of the Michelin guide are relevant today. First, for people today, print guides are associated with a cost, and online guides generally are free. The Michelin guide was free, too, for its first twenty years until, as legend has it, one of the Michelin brothers discovered that the guidebooks were being used as wedges under uneven table legs. Its cost was offset in part by advertisements for Michelin products (such as brake shoes) and for automobiles, garages, and hotels. In an effort at impartiality, Michelin removed hotel advertisements in 1908 (an early lesson that is not being followed today given the recent spate of lawsuits about biased and fake reviews). Second, even Michelin, now known for its discriminating but anonymous "inspectors," used public input. Early guides solicited information from hotels and garages about their establishments. They also welcomed feedback from the public about the accuracy of the guide, including whether the prices listed in the guide reflected actual costs and whether the hotel had bedbugs. Indeed, an early advertisement for the guide portrayed an unhappy wedding night when a hapless couple, without a Michelin guide, made the mistake of staying at a hotel that was ridden with bedbugs.[15] A recent resurgence of bedbugs has made them again a (contentious) topic of (online) reviews.

The Engineers

Much like accusations about hotel bedbugs, digital cameras can be a surprisingly contentious subject. Digital photographers can be argumentative and loyal to their brand—the feud between Canon and Nikon "fanboys" is infamous. The site DPReview, now owned by Amazon, offers forums, ratings, rankings, and reviews. Carefully reading a camera review is not to be undertaken lightly: a review can run over twenty pages with thousands of words and dozens of charts, tables, and sample images and movies. To justify its assessment to "pixel-peeping" skeptics (those who zoom into images and compare them at the pixel level), the site provides a "side-by-side camera comparison" that compares the features and sample images of different cameras.

This approach to review coincides with the early twentieth-century movement of engineers who applied scientific methods, rigorous standards, and progressive planning to economic and social reform. The mechanical engineer Frederick Winslow Taylor famously studied the motions of men who were handling pig iron and recommended ways to minimize their wasted movements. Stuart Chase, educated as an engineer and accountant at MIT and Harvard, was cut of a similar cloth, but focused on the inefficiencies of the larger American economy. Chase's accessible and widely read criticisms grew out of an experience that he had working at the Federal Trade Commission. During World War I, he was part of an investigation of the meat-packing industry. Frustrated by the experience, he concluded that the regulation of the industry was "nothing more than a comedy"—a sentiment that contributed to his dismissal in 1921.[16] However, he did not lose his technocratic ideals. Instead, his path was set upon a socialist cant, premised upon the idea that waste in industry and government could be judiciously pruned so that consumers could enjoy a basic level of security and satisfaction.

Chase's colleague, F. J. Schlink, was a mechanical engineer who also had worked for the U.S. government. His time at the Bureau of Standards, at the quality control departments of Firestone Tire and Bell Laboratories, and at the American Standards Association, imparted a passion for planning and study. In 1927, Chase and Schlink collaborated on a publication that likened the dizzying array of product claims to the outlandishness of Alice's Wonderland: "Why do you buy the tooth paste you are using—what do you know about its relative merit compared with other tooth pastes—do you know if it has, beyond a pleasant taste, any merit at all?" Every toothpaste and every other product claimed itself to be the best and greatest: "We are all Alices in a Wonderland of conflicting claims, bright promises, fancy packages, soaring words, and almost impenetrable ignorance."[17] Their book, *Your Money's Worth: A Study in the Waste of the Consumer's Dollar*, is a rousing argument against manipulative advertising. In its stead, they advocated that claims should be verified against quality standards. Noting that the federal Bureau of Standards spends $2 million on tests annually but likely saves the government $100 million every year, they ask why similar efforts should not be made on behalf of the American consumer. For instance, given that the tire industry had "voiced a warning that tires were being made to last too long for healthy business," they imagined conducting experiments in which tires and automobiles could be tested for longevity and safety.[18]

Their book was a success and brought much attention to the small "consumers' club" that Schlink maintained in White Plains, New York. In 1929, Schlink and Arthur Kallet, the club's secretary (also an engineer and MIT alumnus), founded Consumer Research to work on a national scale. In 1933, they published *One Hundred Million Guinea Pigs: Dangers in Everyday Foods, Drugs, and Cosmetics*. The American consumer was no longer characterized as an "Alice in Wonderland." The country's 100 million residents were test subjects for misleading but well-advertised, sometimes useless, and even dangerous products. Did readers realize that the toothpaste that they brushed their teeth with "contains enough poison if eaten, to kill three people; that, in fact, a German army officer committed suicide by eating a tubeful of this particular toothpaste?" Just as the criticisms found in *Your Money's Worth* are comprehensible a hundred years later, the critiques of *Guinea Pigs* could be taken from today's newspapers' discussions of sweeteners, additives, and the transformation of food into "borderline food substances."[19] Although Schlink was at the forefront of progressive consumer interests, he was not as sympathetic to socialist concerns as his former colleague Stuart Chase was. In 1936, when three of his employees formed a union, Schlink fired them and acted forcefully against the subsequent strike, which he thought was "an unholy alliance" of strikers and "capitalist advertisers" against consumers.[20] The strikers eventually formed the Consumer Union, which became the publisher of the *Consumer Reports* that we know today.

Since its outset, *Consumer Reports* has refused advertising and free samples from manufacturers. (This is not something all blog reviewers can say today.) It currently reports that it has 157 shoppers in 30 states with a testing budget of approximately $20 million. In 2002, it tested 1,863 products, including cars. Given its engineering legacy, *Consumer Reports* continues to focus on rigorous empirical testing of products. Its Website describes efforts to be as statistically accurate and sound as possible. For instance, in a 1999 study, the person responsible for gathering nine thousand condoms did not limit himself to clinics and pharmacies: "When it came time to finding condoms sold in vending machines, the only place he could find them were nightclubs. He spent many an hour self-consciously feeding coins into the machines located in the nightclubs' men's rooms. But he was successful in getting the sample we needed!"[21] This dogged, careful, and analytic approach to commenting about products continues to today.

The Likers

Among his many accomplishments, Kevin Kelly is founding editor of *Wired* magazine and the cofounder of the WELL (Whole Earth 'Lectronic Link), a seminal bulletin board system. He also began *Cool Tools*, a blog for tools that "really work." At his core, Kelly is a *liker*: a tool guru with an enthusiasm for sharing recommendations about stuff, especially items that are "tried and true." Almost anything is within scope, be it a literal tool, a kitchen gadget, or a useful book. Tools can be "old or new as long as they are wonderful." The philosophy of the site is to "post things we like and ignore the rest," and it asks readers to "tell us what you love."[22] *Cool Tools* started as an email list in 2000 but migrated to the Web in April 2003 with a post about a keylike miniknife that might pass through airport security. Unlike many other sites that were launched in blogging's early days, this one is an extension of a comment culture whose roots stretch back to the 1960s.

In 1966, twenty-eight-year-old Stewart Brand was at the beach, tripping on LSD and gazing at San Francisco's skyline, when he noted the slight curve of the horizon and mused that if he ascended, the curve of the earth would become more pronounced until he could see the whole of the earth. Such a perspective could be the jolt that people needed to appreciate that planet Earth was "complete, tiny, adrift." After gaining this insight, people would never "perceive things the same way" and would get on with the business of "getting civilization right." Given the frenzied activity of NASA and the Soviets, why had we not seen a picture of the earth yet? The next morning he began printing buttons and posters with that very question.[23] A couple years later, the *Apollo 8* moon mission delivered the photograph, and Brand's *Whole Earth Catalog*, published regularly between 1968 and 1972 and intermittently thereafter, featured the image of the blue and green marble on its cover.

The stated purpose of the *Whole Earth Catalog* reads like a manifesto: "We are as gods and might as well get used to it." This power arises from the ability of the "individual to conduct his own education, find his own inspiration, shape his own environment, and share his adventure with whoever is interested." Brand's larger take on the world was quite unlike the East Coast accountants and engineers. He trained as a biologist at Stanford University and became an entrepreneurial hippie who, among many things, organized the Trips rock music festival. His enthusiastic vision of sharing and human progress was best represented in the *Whole Earth Catalog*—a few editions of which were edited by *Cool Tools*' Kevin Kelly. The catalog sought "tools that aided this process" of human advancement and were useful, furthered self-sufficiency, provided good value, were little known, but easily purchased by mail. Computers, too, could be powerful tools. As they became "faster, smarter, smaller and cheaper," they shifted the balance from the estrangement of institutional computing to the empowerment of personal computing. There was one problem: "For new computer users these days the most daunting task is not learning how to use the machine but shopping."[24] We could become gods, if we made the right purchases.

Personal computers also could be networked, providing a new way for people to communicate with each other and build community around the Whole Earth ethos. In 1984, Brand began publishing the *Whole Earth Software Catalog*, and in 1985, with Kelly's help, he began the WELL bulletin board system in San Francisco. Communication scholar Fred Turner has argued that much of the Internet's culture is rooted in this West Coast movement from "counterculture to cyberculture." In a 1995 essay for *Time* magazine, Brand himself wrote that "We Owe It All to the Hippies": they provided "the philosophical foundations of not only the leaderless Internet but also the entire personal-computer revolution."[25] LSD, geodesic domes, blue boxes (for phone phreaking), and the personal computer were tools of personal empowerment. Awareness and knowledge about all of this was shared, and this exchange was further amplified and decentralized when it went online at the WELL.[26]

Echoes of this ethos also can be seen in the career of Mark Frauenfelder, cofounder of *Boing Boing*, a prominent blog that has struggled with comment (as discussed in chapter 1). The blog's description of itself as a "directory of wonderful things" reflects that its print predecessor (a zine of the same name) was inspired by the *Whole Earth Catalog*. Frauenfelder first went online by way of the WELL and worked with Kelly at the launch of *Wired*.[27] In the new millennium, the Whole Earth ethos, complemented by the do-it-yourself (DIY) ethic of zines and the skills of hobbyists and hackers, has resurfaced with the ascent of the *maker movement*. Frauenfelder has served as the editor-in-chief of *Make* magazine and in 2013 took the same position for a publishing collaboration with Kelly called Cool Tools Lab. Its first product was a printed book: a "curated selection of the best tools available for individuals and small groups" that was based on ten years of postings from Kelly's blog. In the promotional video for the book, Kelly noted that "pages of the *Whole Earth Catalog* were homemade and very personal, filled with deep enthusiasm and amateur obsession about both old and new ways of doing things. You could learn how to start raising bees or begin blacksmithing.... I began carrying on this tradition in a blog called *Cool Tools*."[28]

The Whole Earth ethos, from the original *Catalog* through *Boing Boing* and *Cool Tools*, exemplifies the idea that sharing one's reviews, *likes,* and *+1s* can be a personal offering that reflects enthusiasms and experiences for the betterment of other people—even if it is about an egg timer, a dog canteen, or a pencil sharpener.

The Crowd

The Zagat review for Veggie Galaxy, a restaurant in my neighborhood, reads: "'Diner classics are reimagined' in 'fabulous,' 'stereotype-defying' vegetarian and vegan guise ('you'll totally forget you're not eating meat') at this 'retro' joint near Central Square Theater; 'courteous' service and 'cheap' tabs round out the 'wonderful concept.'"[29] This Frankenstein sentence is cobbled together from the disparate reviews of ordinary people. In its "About Us" page, Zagat traces the origins of this approach to a 1979 dinner party conversation about unreliable restaurant reviews: "It was at that moment Tim suggested taking a survey of their friends. This led to 200 amateur critics rating and reviewing 100 top restaurants for food, décor, service, and cost. The results, printed on legal-sized paper, were an instant success with copies being scooped up all over town." Zagat now claims to be the "world's leading consumer survey–based leisure information source." As evidence of the value of such commentary, Google acquired Zagat in 2011 for $150 million (and in the summer of 2013 launched a new Zagat site and app that made ratings and reviews freely available online). A Google executive drew a connection between Zagat's origins and user-generated content of today:

Their surveys may be one of the earliest forms of UGC (user-generated content)— gathering restaurant recommendations from friends, computing and distributing ratings before the Internet as we know it today even existed. Their iconic pocket-sized guides with paragraphs summarizing and "snippeting" sentiment were "mobile" before "mobile" involved electronics. Today, Zagat provides people with a democratized, authentic and comprehensive view of where to eat, drink, stay, shop and play worldwide based on millions of reviews and ratings.[30]

Although Michelin and other guides have always employed user input, with Zagat the input *was* the content. However, Zagat's "user-generated content" was different from that of sites like Yelp and Amazon. Zagat staff selected and combined the pithy anonymized excerpts. At Web 2.0 sites, one can see reviews intact and reviewers are identifiable. (However, as I will discuss in the next chapter, the relative prominence of positive or negative reviews is controlled and manipulated by Websites.) This idea that aggregations of decentralized and democratic opinion can be a valuable form of information is often spoken of as the "wisdom of the crowds." Like *UGC* and *Web 2.0*, this term and its cousins *crowdsourcing* and *collective intelligence* are frequently misunderstood. And such buzzwords are not always misunderstood in the same way. Some have little substance, and others have substance that is sometimes forgotten.

The term *Web 2.0* is attributed to a 2004 conversation on the naming of a conference about the reemergence of online commerce after the collapse of the 1990s Internet bubble. Tim O'Reilly, technology publisher, wrote that chief among Web 2.0's "rules for success" is to "Build applications that harness network effects to get better the more people use them. (This is what I've elsewhere called 'harnessing collective intelligence.')"[31] Additionally, the popularization of collective intelligence can be traced back to two men who were associated with the WELL—the Whole Earth 'Lectronic Link. In the 1990s, chaos and complexity theory were hot topics that Kevin Kelly popularized with his book *Out of Control: The New Biology of Machines, Social Systems, and the Economic World.*[32] Kelly showed how order can emerge from seeming chaos: how the beautiful midair choreography of a flock of birds arises when many individuals follow simple rules of interaction. This "new biology" was mostly gleaned from and applied to the natural world, but Kelly also posited it as a theory of social organization and intelligence via the notion of the "hive mind." This idea persisted into the new millennium, when varied new media-related phenomena required explanation. In 2002, Howard Rheingold, another famous WELL member who had previously authored a seminal and popular treatment of virtual communities, published *Smart*

Mobs.[33] In this latter book, Rheingold argued that new forms of emergent social interaction would result from mobile telephones, pervasive computing, location-based services, and wearable computers. Two years later, in *The Wisdom of Crowds*, James Surowiecki made a similar argument, but instead of focusing on the novelty of technological trends, he engaged directly with the social science of group behavior and decision making.[34] Surowiecki argued that groups of people can make good decisions when there is diversity, independence, and decentralization of opinion and when that information is appropriately aggregated. An open question (to which I will return) is whether the sites that feature user reviews (like Yelp and Google) are sufficiently impartial (with respect to their own interests) and fair (about those that they review) to qualify as providing "wise," or at least useful, comment that informs.

The Critic

This brief historical excursion has been a first step on an expedition to the bottom half of the Web. But one last archetype should not be missed: the critic. In the age of the Web, the question of who gets to be a critic has been contentious. This can be seen in James Berardinelli's career as a film critic, which spans a historical inversion. In three moments separated by roughly seven years each, this amateur reviewer and his peers were portrayed as a novelty, an invasion, and the death of "serious" criticism.

In 1997, the *Los Angeles Times* profiled the then twenty-nine-year-old Berardinelli, author of over twelve hundred film reviews, in an article titled "In Online World, Everyone Can Be a Critic."[35] Berardinelli was noted as one of the best online amateur reviewers. He posted his first review to Usenet in 1992 and on his Website ReelViews in 1996. Berardinelli related his efforts explicitly to the love of movies, a passion demonstrated by the many hours he spent on reviewing in addition to his continued work as an engineer. Reviewing itself provided little monetary compensation; instead he used his day-job to provide "enough money to pay the mortgage, keep up my home theater, finance film festival trips, and buy the 20 gallons of gasoline I need each week to attend screenings."[36] Yet, what he hasn't gained in coin, he has gained in personal satisfaction. The site Screen Junkies lists him as one of the ten most famous movie critics, along with Rex Reed, Roger Ebert, and Pauline Kael. His online reviews have been collected in print books, one of which is favorably introduced by Roger Ebert. He even met his wife through his Website.

Yet, does his passion for movies necessarily make him a "film critic"?

> When I first started reviewing in 1992, I rigorously avoided the term "film critic" because it was a label I didn't feel I had earned. I referred to myself as a "film reviewer." It wasn't until the late '90s, after the website was on-line and I had 1000 reviews to my name, that I became comfortable with the "film critic" label. I am a populist critic, which means I write for the masses. That's not to say I am incapable of writing deeper, more literate essays, but the general purpose of a 700- to 1000-word review is to provide an informed opinion about a movie. My goal with a review is threefold: provide my opinion and explain it, present enough information so that someone reading the review will be able to make a determination about whether they might like it (irrespective of whether or not I did), and offer some insight that those who have seen the movie may find interesting. I have some longer pieces on the website for older movies that can run up to 2000 words. Those typically contain more critical analysis than the "regular" reviews.[37]

The composer and lyricist Stephen Sondheim makes a similar, somewhat more nuanced distinction. Where Berardinelli sees the label *critic* as a badge of distinction to be earned with practice, Sondheim makes a functional distinction: "Reviewers are reporters; their function is to describe and evaluate, on first encounter, a specific event," and because they often work on a deadline, some become blandly enthusiastic or cynically jaded. A critic, on the other hand, also describes and evaluates, "but from a loftier perspective" provided by time and distance: "That loftiness sometimes leads them to promote themselves rather than the object of their affection, but loftiness is what the readers look to them for, and often with rewarding results. What readers look for in a reviewer is immediate guidance." Hence, a reviewer does not require any special knowledge. Echoing the motive of editors of the *Monthly Review* from the eighteenth century, Sondheim writes that "People read reviews to decide whether they should spend a considerable sum of money to see for themselves the subject under the microscope."[38] Although Berardinelli took writing classes in college, has read countless books on film and its history, and has attended many symposiums, it is his "great love and appreciation for movies" that characterizes his efforts. He believes that requiring a formal film education to write movie reviews is "the height of arrogance": "One of the great things about movies is that almost everyone has an opinion, and it's rare that any two will be the same. Film criticism is not surgery— you don't need a degree to be an effective practitioner of it."[39]

By 2004, the proliferation of online film reviews prompted *Wired* magazine to announce an "Invasion of the Web Film Critics."[40] It had become possible for those writing for online publications to be accredited by the studios, meaning that they could see advance screenings. (Before Berardinelli was accredited, he attended many opening nights to be current.) Some of these writers even earn a living by writing for online publications like *Salon* and *Slant*. Moreover, few print reviewers could long ignore the need for their reviews to be available to and engage an online audience. Aggregators like Rotten Tomatoes (1999) and Metacritic (2001) were new forces to be reckoned with and evidence of the logic of the crowd and the recurrent theme that comment often prompts new comment and new means of managing it.

What was happening in film was not unique. The arrival and "invasion" of online reviewers was manifest across media, including music and literature. The critics who evaluate "from a loftier perspective" (according to Sondheim, those who read "to learn something about the cultural landscape"), naturally question their own role in the new landscape. And they have always done so: Michel Foucault, T. S. Eliot, Oscar Wilde, and Walt Whitman each attempted to define the role of the critic. In 1960, Alfred Kazin did so in an essay in the *New York Times Book Review* titled "The Function of Criticism Today," and in 2010, the *Review* asked six "accomplished critics" to examine current criticism by reflecting on Kazin's essay. The *Review's* editors noted that "We live in the age of opinion—offered instantly, effusively and in increasingly strident tones." Much of this goes by the name of criticism, but "where does it leave the serious critic?"[41]

Stephen Burn, scholar and author, wrote that "While Kazin could complain in 1960 that 'the audience doesn't know what it wants,' with the advent of Amazon reviews and other rating sites the audience is abundantly vocal." Indeed, "the audience now talks to itself.... The age of evaluation, of the Olympian critic as cultural arbiter, is over." Yet the critic can still provide a valuable function by exhuming a work's context and placing it within a larger frame. Writers Katie Roiphe and Sam Anderson both argue that critics can distinguish themselves by writing well. For Anderson, the role of the critic is to amplify the conversation: "we make the whispered parts of it audible; we translate the coded parts into everyday language." For Roiphe, critics "have always been a grandstanding, depressive and histrionic bunch," but if they wish to compete with "the seductions of Facebook ... [and with] every bright thing that flies to the surface of the iPhone," they must write beautifully. Only by exemplifying grace in thought and writing can they have any authority to separate the talent from the transitory: "There is so much noise and screen clutter, there are so many Amazon reviewers and bloggers clamoring for attention, so many opinions and bitter misspelled rages, so much fawning ungrammatical love spewed into the ether, that the role of the true critic is actually quite simple: to write on a different level, to pay attention to the elements of style."[42]

Unboxing

A function of comment is to inform—to share our thoughts for the benefit of others. This motive was apparent in the earliest days of digital communication. Internet pioneer Vint Cerf noted that "when e-mail showed up in 1971 on the ARPANET, we discovered instantly that e-mails were a social network phenomenon." The evidence was the quick appearance of two email lists that were dedicated to "book reports and restaurant reviews"—the SciFi-Lovers and Yum-Yum lists.[43] And many models of review that are now common on the Web precede the digital age itself. Stars arose a century ago to discern relative worth, engineers provided detailed comparative analysis, and likers shared recommendations that were rooted in love and experience. The crowd shared its particular, peculiar kind of wisdom, and the critic highlighted and connected with analysis and insight. All that has gone before is present on the Web—and more. Each of these types of informing comment existed before the twenty-first century but never in such number. Nor were they ever so easily accessible as a barcode and a smartphone. Also, there are now genres of review that include and amplify earlier forms into something new. The Wikipedia article on "Unboxing" likens it to "geek porn" and describes it as a video of the "unpacking of new products, especially high tech consumer products." The earliest instance of the term appears to refer to a 2006 YouTube video of a Nokia E61 smartphone. Yet an unboxing is not a comparative analysis or an expert review. In the age of the Web, where both gadget lust and conspicuous consumption operate on the thin edge of time, *unboxing* is a novel genre and new ritual. For some, it is even a way to make money, as reviewers buy (and later return) products solely to unbox them online.[44]

Interestingly, these videos are somewhat gendered as well. Unboxing videos are mostly by men about gadgets received by mail; *haul* videos are more often the results of women's shopping trips to local stores. Additionally, different types of products have their own subgenres of review. In 2004, a physical therapist told me that because my shoulders were askew, I should stop using a messenger-style bag over one shoulder and instead use a backpack that evenly distributes weight. Fortunately, I found one that suited me at a thrift shop. Unfortunately, almost ten years later, it was disintegrating and novel finds at thrift stores are not repeatable. So I turned to the Web. Product reviews are numerous and popular on YouTube—high-tech unboxings are only the tip of the iceberg. One can find video reviews for silly putty, the egg genie, a pancake pen, and the double bullet (a sex toy).

There are also the reviews from the "doomsday prepping" survivalist community. There are an estimated three million "preppers" in America and some spend hundreds of thousands of dollars on their bunkers and gear. (They even have their own dating sites.) At YouTube, prepper reviewers are typically white Christian men who are concerned with an over-reaching big government, gun rights, and the collapse of civil society. Their slogan is "pray for the best, prepare for the worst." As one blog posting noted, these folk are "completely obsessed with both gear and the idea of self-sufficiency. They prize durability and functionality in a product because their fervency makes them believe their lives will depend on it."[45]

Many of these reviews are for Maxpedition products, a reputable but expensive brand that initially sold to the military, law enforcement, and emergency responders. Its market expanded when the FR-1 medical pouch was adopted by survivalists. A rural survivalist's bag will likely include maps, cash, flashlights, a handgun, hand sanitizer, a compass, a GPS navigator, knives, toothpaste, bandages, food bars, water filters, antibiotic ointment, parachute cord, and a battery charger for gadgets— among many other things. (Flashlights and knives are fetishized objects that garner many reviews.) Like members of other subcultures, survivalists have their own lingo. For instance, a "bug-out bag" is a prepacked bag that can be grabbed out of a closet or car trunk and help people survive for seventy-two hours after a disaster. (Discussions about the ideal contents of a bug-out bag are extensive.) An "EDC" is an "everyday carry bag." A video "load out" is much like an unboxing video, except that as the reviewer unpacks the bag, he discusses his loading strategy and the merits of each item. Many reviewers have military experience or have adopted military vocabulary and speak of PALS webbing and MOLLE-compatible attachments.

The backpack that I purchased, the Maxpedition Pygmy Falcon-II, has dozens of YouTube reviews, some of which are fifteen minutes long. In my favorite review and "load out," a young man begins by testing the bag's stability while he attacks a martial arts dummy and then jumps rope. He admits that jumping was not a good idea because the pack comes down when he goes up. As he unpacks, he finds the Bible, the Declaration of Independence, and an anti-Obama tract within its pockets.[46] These materials are common in prepper reviews and are reminiscent of the Crystal champagne that rappers often have chilling in their refrigerators on MTV's *Cribs*. Reviewers take their task seriously, though sometimes one cannot help but laugh at the bravado. In one odd juxtaposition, an Amazon reviewer of the Falcon-II reports that "I bought this for my 5th grader [and it] works very well." Additionally, the bag can fit an M4 assault rifle, although "it does not get a five star because the drag handle is small if you needed to drag a wounded team mate while wearing gloves and under fire."[47] (I wonder, how often will his fifth grader need to carry an M4 or drag a wounded teammate?) One can even find some humor, in which rugged survivalism is replaced with domesticity. In one case, a "go-bag" became a favorite diaper bag because the adjustable straps could be quickly fit to the father or mother and the main compartment fit wipes, a full pack of diapers, and other miscellany. Additionally, "there are 2 water bottle holders, which is perfect for carrying a water bottle for you, and a sippy cup for your kid."[48]

Despite the relative novelty of the unboxing and haul videos, many insights learned from the past can be applied to online comment today. For instance, comment begets more comment, as was seen in early literary reviews and the glut of social media buttons today. Also, the tensions between public input and expert opinion preceded and continue into the digital age, as do arguments about who can claim to be a critic. Most important, the historical proliferation of comment accompanied an increase in consumerism, as seen in the story of the Michelin stars.

This point seems especially salient today. Although many types of online informing comment have historical antecedents, the scale and pervasiveness of comment today are remarkable, and much comment is related to the consumption of goods and services. Online comment is worth billions of dollars and subject to much manipulation—the topic of the next chapter.

3

Manipulated: "Which Ice Cube Is the Best?"

So I've got all these apps on my phone to let me rate restaurants, hotels, food. But why have all these apps when I can have just one that lets me rate everything?

Take for example this glass of water, which ice cube is the best? According to Jotly: that one, it's melting the slowest. Helpful! ...

It just makes sense. With the Jotly you can rate anything. You don't need all of those other apps. I mean they all got pretty low ratings anyway. So just delete them.

—Promo Video for Jotly

In a 2011 video for the fictional Jotly app, an enthusiastic user descends a playground slide, and the app informs him that "5 others have slid here. 10% go faster than you. Share on Twitter!" With a frown, he comments that "this slide is not very exciting" and gives it a "D."[1] This video parodied a proliferation of rating services and applications. In fact, Jotly appeared at roughly the same time as the genuine rate-everything apps Oink and Stamped. Yet, while these were soon discontinued, an actual Jotly app was released with its sardonic edge intact: "Your life is exciting and worth sharing: everything with everyone! Everyone cares about everything you do. Now you can rate your entire life and share the experience."[2]

These apps hint at the immense value of users' online comments, including reviews and ratings. In 2013, Amazon purchased Goodreads, the book review and discussion site, for $150 million. Within four years of its founding, TripAdvisor, a travel review site, was acquired by Expedia for $200 million. This 2004 success paled in comparison to TripAdvisor's 2011 initial public offering as a separate company valued at over $3 billion. Google purchased the restaurant rating guide Zagat in 2011 for $150 million. However, Zagat may have been Google's second choice. It is rumored Google offered Yelp $500 million but Yelp walked away, perhaps embittered by instances of Yelp reviews appearing without attribution in Google Places. In 2012, Google continued to feed its appetite for reviews by purchasing Frommer's Travel for $22 million. In 2013, Google sold the Frommer's publishing brand back to its octogenarian founder Arthur Frommer: it had the data that it wanted and few plans to do anything further with the brand.[3]

User comments, ratings, and reviews are valuable because, in economic terms, they address the marketplace problem of *information asymmetry*. The classic example is a potential buyer's fear that a car might be a "lemon." The seller can signal quality by offering a warranty, noting resale value, or (if it is available) providing the car's maintenance history and mileage. Because some of these things can be manipulated, such as turning back a used car's odometer, additional sources of opinion are sought, including informal word of mouth (does your friend like the one she bought?), traditional reviews (what did *Consumer Reports* conclude?), or online reviews (what do users on Yelp say?).

The information asymmetry is especially problematic when we buy things sight unseen. In the late 1990s, the main concern that people had with online retail was the trustworthiness of the merchant: was the retailer just a scammer with a Webpage? In response, online merchants decorated their pages with assurance seals that vouched for their legitimacy. But this told consumers little about the quality of a particular product. One way to deal with concerns about quality is to have a friendly return policy. Because online shoe stores deal with the fact that buyers never know how well the shoe fits until worn (researchers call this an "experience product"), Zappos (and others) provide convenient and free returns. Some Zappos devotees think nothing of buying three different sizes of a shoe and returning the two that do not fit.

It would be ideal if people could avoid making poor purchases in the first place. One approach is for the merchant to select the best products, such as the stylist who curates Shoe Dazzle's collection. The opposite approach is to sell everything, and let consumers sort it out. Zappos customers review and rate its ninety thousand styles of shoes. Amazon, which acquired Zappos in 2009, sells billions of products to its over 200 million customers. In 2000, Jeff Bezos, founder of Amazon, said, "We want to make every book available—the good, the bad, and the ugly. When you're doing that, you actually have an obligation—for going to make the shopping environment one that's actually conducive to shopping—to let truth loose. That's what we try to do with the customer reviews."[4] Today Amazon sells more than books (in 2010, media sales were surpassed by sales of other products), but the philosophy of letting "truth loose" remains the same.

Amazon and others have "let truth loose" with user comment, but that truth is being overtaken by fakery and manipulation. TripAdvisor's extraordinary growth in value has been accompanied by increasing stories of abuse: of proprietors who write their own positive reviews, slam competitors, or pay others to do so. In 2013, a hotel chain executive was writing reviews under a pseudonym, but when he did so using TripAdvisor's Facebook app, his real name, photo, and location were exposed publicly. Was this fakery the exception or the norm? Who is writing false reviews and why? And what is being done to counteract it? In this chapter, I address these questions and argue that replacing strangers' recommendations with friends' endorsements will not solve the problem. Sites that take advantage of our social networks for commercial interests hasten a world in which we are all increasingly tempted to take advantage of each other.

Research on Reviews

Marketing consultants have issued dozens of reports that document consumers' reliance on online user reviews—and these firms have a marketing service they want to sell you. Fortunately, the nonpartisan and nonprofit Pew Research Center offers some compelling findings. In a 2010 telephone survey of 3,001 adults, 58% reported that they have gone online to research products and services, and 24% said that they have posted comments or reviews about their purchases. (Among Internet users, this is 78% and 32%, respectively.) In fact, 21% of respondents report they had researched a product the day before. In a 2011 Pew survey, 55% of adults said that they seek news and information about local restaurants, bars, and clubs, and 51% of those turn to the Internet for information.[5]

This use of reviews by consumers affects a merchant's bottom line. In one early study of eBay, researchers offered identical vintage postcards as a reputable merchant and as a newcomer, and the established merchant's postcards sold for 7.6% more than the newcomer's. More recently, a study of digital cameras listed on the price-comparison site NexTag found that a camera's poor rating was associated with a greater discount from the manufacturer's suggested retail price. Another early study compared how book reviews at Amazon and Barnes & Noble affected sales rankings. Because it can be unclear whether better reviews lead to better sales or if both are simply the result of a better product, using the same product on two different sites helps researchers to focus on the effects of the reviews. In this study, the conclusion was that better reviews were associated with better sales.[6] When it comes to dining, clever researchers noted that Yelp rounds off the overall star rating to the nearest half star. So a restaurant with an average of 3.24 is rounded to 3.0 stars, and one with an average of 3.26 is rounded to 3.5 stars. One researcher used discontinuities around these thresholds to estimate that a one-star increase leads to 5% to 9% more revenue. Other researchers looked at how often a restaurant's tables sold out during the busiest dining hours (using a restaurant reservation site's database) and found that crossing the 3.5-star threshold increased a restaurant's chance of filling all of its tables by another nineteen percentage points. Good ratings and reviews have also been shown to increase the sales of hotel rooms and video games. And even when the *valence* of a review (that is, whether it is positive or negative) does not drive sales, such as in one study of movie ticket sales, the number of comments does: increased awareness seemingly drives greater ticket sales.[7]

There are hundreds of similar research reports on user ratings and reviews. Scholars study the effects of reviews relative to their valence, variance, volume, and helpfulness. They also analyze reviewer characteristics (familiarity and expertise), reader demographics, and the categories and features of the products. From all of this, some high-level conclusions can be drawn. First, ratings and reviews tend to be more significant for niche or relatively unknown products. Second, reviews tend to follow a "J-shape" distribution because people are most likely to participate when they have a strong positive or negative experience. Imagine a very wide *J* with some negative ratings on the left, a dip in moderate ratings in the middle, followed by a peak of positive ratings on the right. Third, while negative reviews do seem to have a greater impact on purchasing decisions, a few negative ratings are not necessarily detrimental: they might prompt consumers to more carefully consider the reviews or have greater confidence when they can see the balance of opinion. Finally, the sentiments expressed in the text of a review are often more persuasive than its corresponding rating.[8]

This last finding is likely related to the fact that people are more suspect of a numerical rating than a textual review. But people are fooled by text as well; some realized the extent of this problem only when they noticed duplicated reviews. Such was the case for Trevor Pinch, a sociology professor at Cornell University, who noted that reviews of a book that he coauthored about the history of electronic music were similar to those of another book. Intrigued by the incident, Pinch and his colleague Shay David analyzed fifty thousand user reviews of over ten thousand books and CDs and identified three likely motivations. First, different reviewers copied positive reviews from other items to their own to promote sales. Second, the same author might post the same review on multiple items to "promote a specific product, agenda, or opinion." For instance, a creationist might post the same critical review across multiple books on evolution. She might do this from the same account or through multiple accounts, known as *sockpuppets*. Finally, reviewers might plagiarize their own reviews across products to increase their review count and credibility (although, as I discuss below, some top reviewers' numbers lead to incredulity). This analysis allowed David and Pinch to conclude "that about one percent of all review data is duplicated, verbatim or with variations."[9]

The use of duplicate reviews also has given researchers an interesting tool for understanding review manipulation. Computer scientists Nitin Jindal and Bing Liu performed an analysis of 5.8 million reviews from 2.14 million reviewers. Because they knew that duplicates were fake, they used these reviews to train a computer model on what fake reviews look like. They asked a computer to use duplicates to learn how to characterize fakes across thirty-five different factors, including the number of comments on a review, a review's rated helpfulness, the length of a review's title and text, and the ratio of positive to negative words (*great* versus *horrible*). They also included whether "a bad review was written just after the first good review of the product and vice versa" (manipulators often follow an authentic review with its opposite to maintain a product's rating). They found that fakery was widespread and that helpfulness ratings were of little help in discerning it. In a 2012 *New York Times* story about "The Best Book Reviews Money Can Buy," Liu was quoted as saying that "about one-third of all consumer reviews on the Internet are fake."[10] Nan Hu, a business school professor, has also focused on how a positive review often follows a negative one. Hu and his colleagues studied 610,713 reviews of 4,490 books (that each had at least thirty reviews) and found that the probability of seeing a positive review after a negative one was 72%, almost 2.6 times the probability of a negative review following a negative review.[11] Clearly, this is suspect, leading them to conclude that 10.3% of the products in their study were subject to manipulation. Researchers who analyzed Boston restaurant reviews found that 16% of the reviews were filtered as possibly fraudulent by Yelp, and these reviews tended to hold more extreme views, are written by reviewers with less established reputations, and focused on restaurants that were facing increased competition.[12] Fake reviews are common, and many expect that they will become more so.

Of Spam, Fakes, and Sockpuppets

Online comments often exhibit what linguist Donna Gibbs calls *cyberlanguage*, "with its own brand of quirky logic, which evolves with unprecedented speed and variety and is heavily dependent on ingenuity and humor."[13] An example of this is the word *spam* (much to the displeasure of the Hormel Foods Corporation, maker of the canned-meat product of the same name). The origin of the online usage is a Monty Python television sketch in which Spam was inescapably (and frustratingly) included in nearly every dish on a restaurant menu. So *spam* became a word for unwanted but inescapable messages. It has also been (inappropriately) used to describe fake ratings and reviews. Both unwanted email and fake reviews exploit vulnerable computers in similar ways and are contracted for on illicit markets, but they are not the same. Spam might be considered a fake ham, but few people would confuse the two. In the Monty Python sketch, it is clear that restaurant customers will be getting "Spam bacon sausage and Spam." In South Korea, in fact, Spam is not poor man's ham but an expensive delicacy. Accordingly, I prefer to speak of unwanted (but possibly truthful) messages as *spam* and manipulated reviews as *fakes*.

Like the word *spam*, the term *sockpuppet* entered Internet vernacular in the mid-1990s to describe people who create accounts so that they can masquerade as others. These identities can praise or agree with their creator, much as an old sock on my hand can easily be made to agree with me. *Strawman puppets* also can be used to take contrary but poorly argued positions. Taking an example from history, this latter tactic is the likely source of Galileo's troubles with Pope Urban VIII. The seventeenth-century scientist and the Roman Catholic pontiff had begun their relationship amiably, meeting a handful of times during which they spoke of the "two chief world systems"—the Ptolemaic (geocentric) and the Copernican (heliocentric). Urban expressed interest in Galileo's work and welcomed a presentation of the relative merits of both the earth- and sun-centered theories. Galileo did this in his *Dialogues*, a genre that has been used since the ancient Greeks in which characters discuss opposing theories. However, Galileo went beyond an impartial consideration and presented a forceful argument for heliocentrism and included the Pope's words in the mouth of "Simplicio," a geocentric simpleton. In the online vernacular, Simplicio was Galileo's *strawman sockpuppet*, which naturally upset the Pope.

Galileo was not the first or last to create a persona to further his own aims. Yet Galileo, like many authors who have written pseudonymously, was not trying to deceive. The practice of writing fake reviews for one's own benefit seems to have bloomed with the rise of the genre of the review itself. As mentioned in an earlier chapter, eighteenth-century literary and scientific authors were making their living and reputations by competing for the limited money and attention of their readers, both in original works and via reviews of others' original works. Botanists Johan Andreas Murray and Carl Linnaeus wrote fake reviews, with the latter concluding that his *Systema Naturae* was "a masterpiece that can never be read and admired enough."[14] Scientists continue the practice today, creating sockpuppet personas whom they recommend to journal editors as qualified reviewers. On the literary side, Walter Scott was known to review his own work and in 1808 wrote an (anonymous) review of Great Britain's great living poets, including himself as "the minion of modern popularity; for the works of no living, and of few dead authors, have been so widely and so rapidly diffused.... The effect of this extensive popularity has been almost ludicrous."[15] Walt Whitman is known to have falsely boasted about the success of *Leaves of Grass* (1855) and in an anonymous review predicted that "he is to prove either the most lamentable of failures or the most glorious of triumphs, in the known history of literature."[16] Anthony Burgess, author of *A Clockwork Orange*, confessed to his own manipulations, noting that "There is something to be said for allowing a novelist to notice his own novel: he knows its faults better than any casual reader, and he has at least read the book."[17]

Putting a Face to Fakers

Researchers estimate that between 10 to 30 percent of online reviews are fake. The cast of manipulators includes *fakers* (those who deceptively praise their own works or pillory others'), *makers* (those who will do so for a fee), and the *takers* (those who avail themselves of such services).

There are plenty of fakers among authors, especially among amateur and aspiring writers. But even well regarded and successful authors are tempted to buttress their standing and bulldoze others'. In 2004, John Rechy, a prolific novelist with a fair amount of recognition, was found to have granted himself (anonymously) a five-star review on Amazon. The *New York Times* reported that his identity was exposed when a glitch at Amazon revealed its reviewers' identities. Although Amazon discontinued anonymous reviews, the use of pseudonyms continues. Rechy justified his actions as payback for unfavorable reviews, which is consistent with research that shows that a positive review often follows a negative one: "That anybody is allowed to come in and anonymously trash a book to me is absurd. How to strike back? Just go in and rebut every single one of them."[18] British writers of thrillers seem especially prone to talking up their work. Stephen Leather confessed that he used sockpuppets to create a "buzz," and R. J. Ellory was found to be writing rave reviews of his own works and attacking others via a pseudonym.

Even academics, who trade in reputation rather than royalties, have been exposed for fakery. When the popular and award-winning British historian Orlando Figes was exposed for trashing his rivals' books, he first claimed that he had been set up, then blamed his wife, and finally apologized.[19] In the same year, a review of historian Simon Winder's *Germania* was removed from Amazon as an unfair personal attack. When fellow historian Diane Purkiss was revealed to be the reviewer, she posted a new review under her name and explained the circumstances of her first review:

I posted a somewhat critical review of this book about two weeks ago, and the author, Mr Simon Winder, has had it taken down on the grounds that I am an academic using a pseudonym. That's true. I am, but not out of any malfeasance—solely because I would not want to mislead readers into thinking that I as an Oxford don have any high-level expertise in the subject matter of books I read as a general reader.... This was my response not as an expert, but as a reader. As a reader I was bored and irritated.[20]

Winder responded to the review in a comment, claiming her original review "impugned me professionally":

Of course I have no problem with bad reviews. I asked for the review to be removed because I didn't fancy an anonymous accusation like that hanging around online. Her review in its current form is fine and I am grateful to her for making it clear that she was slurring me in a private capacity under a false name, rather than as an Oxford academic.[21]

Besides the financial considerations, what seems to set many on the path of fakerhood is the lack of control over their public identity and the power resulting from manipulating others'. Scott Adams, creator of the popular Dilbert comic series, is known for his provocative blog posts. In one, he wrote that men's best strategy for dealing with women is to treat them like children. Although he said that he was not equating women with children, he also said that "You don't argue with a four-year old about why he shouldn't eat candy for dinner. You don't punch a mentally handicapped guy even if he punches you first. And you don't argue when a women tells you she's only making 80 cents to your dollar. It's the path of least resistance. You save your energy for more important battles." Much discussion and criticism followed, though he was not without his defenders, including the user "plannedchaos." On Reddit and MetaFilter, this user attacked those who were critical of Adams and repeatedly noted that Adams "has a certified genius I.Q." When the account was exposed as Adams's puppet, he wrote on MetaFilter that "I'm sorry I peed in your cesspool. For what it's worth, the smart people were on to me after the first post. That made it funnier."[22] Adams later justified his use of the pseudonym as permitting his speech to be judged on its merits and not harm his business. I do not condone either Purkiss's or Adams's use of pseudonyms, but their examples arise from the blurred boundaries of identity and place, an issue that is examined in subsequent chapters.

Some are contrite when they are exposed and follow the incident with a confession. Many fakers begin by noting that they started by wanting to set something right and then turned to self-puffery and attacks on others. In his last piece for *The Independent*, journalist Johann Hari wrote "a personal apology" for taking liberties with the quotations of his sources and for inappropriate edits to Wikipedia. He wrote that "several years ago I started to notice some things I didn't like in the Wikipedia entry about me, so I took them out." This was followed by removing "nasty passages about people I admire" and factually correcting other biographies: "But in a few instances, I edited the entries of people I had clashed with in ways that were juvenile or malicious: I called one of them anti-Semitic and homophobic, and the other a drunk."[23] Such attacks on biographies are a significant source for the dozens (sometimes hundreds) of sockpuppet investigation that are requested every week.

Fakery is not limited to commercial motives or authors. Fake reviews can be used for ideological purposes, such as to censor a viewpoint or laud a politician. Fans of Michael Jackson organized on Twitter and Facebook to suppress a Jackson biography: they objected to positive reviews, posted negative reviews, and advocated that the book be removed from sale—which it was for a short time. Ideological reviews of a different sort were made during the 2012 U.S. presidential campaigns when Barack Obama visited a Florida pizza parlor. When the owner, a registered Republican, made the most of the visit by endorsing and embracing the president in a bear hug, the owner's Yelp page exploded with hundreds of partisan, nonpizza-related comments. Oddly, one study of a clothing store's Website found that many deceptive reviews were written by loyal customers who posted reviews for products that they did not buy and did not want to see continued. The study's authors suggested that people did this to enhance their standing on the store's site or act as "self-appointed brand managers," noting that the most loyal customers are often the harshest.[24]

Companies and governments take part in fakery as well. *Astroturfing* is the practice of simulating authentic grassroots support for a company or government. The electronics firm Samsung has reportedly paid people to comment in its favor, criticize its competitors, and participate in and hype its smartphone application challenges. Governments use the technique as a form of propaganda by extolling the virtues of those in power, removing negative comments, and attacking critics. The U.S. military uses sophisticated "persona management software" (that is, sockpuppets) to support "classified social media activities outside the U.S., intended to counter violent extremist ideology and enemy propaganda." Other nations have been documented using these techniques on their own populations. China and Russia reportedly hire civilians for the job, and in Thailand, commenting in support of the monarchy is the duty of a unit of uniformed soldiers. In South Korea, the head of the National Intelligence Service and other high-ranking officials have been indicted for their covert Twitter campaign during the 2012 South Korean presidential elections.[25] If something or someone can be applauded or pilloried in a comment—whether a hotel, gadget case, plumber, doctor, singer, or politician—there will be fakes.

The Illicit Markets of Makers and Takers

Fakers today no longer need to do it themselves or go it alone: *takers* can pay for the services of *makers*. In 2010, when Jason Rutherford launched GettingBookReviews, his basic $99 service involved posting a review to the major book sites within seven to ten days. Today, takers can pay much more or much less for this type of service, although today makers tend to be a little more discrete than Rutherford.

At the high end, fakery can be an expensive service that an agency provides as part of a reputation-scrubbing or promotional campaign. Journalist Graeme Wood's introduction to the "world of black-ops reputation management" began with some schadenfreude. Wood's former Harvard classmate was in trouble with the Internal Revenue Service for helping his mother hide the family's fortune (by moving bags of cash around the world). Wood wrote that "as a taxpayer and a jealous prole, I watched his downfall with special interest, and even set up a Google Alert to keep abreast of developments." But the "developments" were press releases and Websites that extolled his classmate's professional and philanthropic successes. These were being orchestrated by a "boutique shop for the online reputations of very wealthy people" that charged as much as $10,000 a month to drown out negative press about its clients.[26]

Fakery can be had on the cheap as well, and most of these ad hoc services are found on sites like Freelancer, Fiverr, and Craigslist. On Craigslist, for instance, those who are willing to write piecemeal reviews are asked: "Are you a savvy writer? Do you like to write book reviews? Are you willing to write a glowing review for a cookbook? Are you an Amazon.com customer or member?" If so, "You will receive a Los Angeles Metro Discount Coupon Card for your participation. Save on pizza, car repairs, etc. $25 Value." Some ads are written by people who are willing to review, such as the "ACTIVE YELP WRITERS" who "WANT TO WRITE REVIEWS 4 YOUR BIZ." They explained that "we aid you obtain a better reputation by acquiring you favorable reviews," which they claimed will lead to more customers. But this ad does not sound as though it was written by a native English speaker. In fact, it probably was created by an overseas "review sweatshop" where workers will never experience the things that they review. Even tech-savvy homeless people can get in on the game with an old laptop and access to wifi and power outlets. In an August 2013 story, unemployed computer technician Jesse Angle estimated that he had earned about $500 via BitCoinGet, a site that pays "every time you watch a video or complete a task"—likely inflating viewership numbers.[27]

Among "made in America" review makers, the going rate seems to be about $5 to $20 dollars per review. Some review takers can require reviewers to visit the establishments and have an existing account on sites like Yelp, Citysearch, and Facebook. One such sophisticated ad even required that the reviewers write a range of reviews (to avoid appearing overly positive) over a period of time (because a rush of positive reviews is suspect) from unique accounts: "Reviews should be very different from each other—ie, one might say 'Item was shipped quickly' and another might say 'A+ great service!!' while another (3-star) might say 'I was satisfied with their customer service', etc."[28]

For some of the most prominent Amazon reviewers, it's not immediately clear what they gain, but there are suspicions. In a 2008 essay in *Slate*, Garth Hallberg noted how pleased he was to receive a five-star review from Grady Harp, Amazon's seventh top reviewer with over 3,500 reviews to his name. Hallberg wrote "Sure, he'd spelled my name wrong, but hadn't he also judged me 'a sensitive observer of human foibles'?" And over one hundred readers clicked that they found the review helpful. However, "after a brief e-mail exchange, my publicist confirmed that she'd solicited Grady Harp's review."[29] Comments on reviews and in discussion boards are full of questions about the legitimacy of top reviewers. Hallberg noted that Amazon reviewers are suspected of receiving free promotional copies or other benefits (including payments). Also, Harp's one hundred votes of helpfulness might indicate the existence of "backscratching" wherein cliques of reviewers rate one another's reviews as helpful. Finally, many wonder how top reviewers can review so many books. Could Harriet Klausner, a woman who dominated Amazon's reviewer rankings for many years, read and review, on average, forty-five books per week for over five years (even if she is a speed-reading insomniac, as she says)? A commenter on the (sarcastically titled) *Harriet Klausner Appreciation Society* blog noted that if she kept the books that she has read, she could fill over 240 bookcases, although it appears she sells some of the books online at half.com. Critics also question the usefulness of top reviews given that the vast majority of those reviews are positive: more than 99 percent of Klausner's reviews are four or five stars relative to the (still high) 80 percent of all reviews.[30]

For fake-review takers, the results can be impressive. Author John Locke was one of the first to sell a million $0.99 self-published ebooks on Amazon. He followed this success with *How I Sold One Million eBooks in Five Months* (the Kindle edition sells for $2.99). However, as "the most helpful critical review" of the book on Amazon noted, there is "a secret he left out": "In an interview with Locke in today's *New York Times*, he admitted that he paid for 300 reviewers to heap praise on his books, a sleazy promotional technique that seems to have worked for him."[31] In the *Times* article, journalist David Streitfeld noted that Locke was confident in his writing but frustrated with his efforts to reach readers via labor-intensive blogging, tweeting, and personalized email. So he made use of Rutherford's GettingBookReviews, starting with purchasing fifty reviews for $1,000 and eventually purchasing three hundred. Locke specified that "If someone doesn't like my book they should feel free to say so" but that reviewers should purchase books from Amazon so they would show as "Amazon verified purchases," lending credibility to reviews and improving the book's Amazon sales rank. Locke has sold millions of inexpensive digital books, and in 2011 he scored a deal with Simon & Schuster to manage the sales and distribution of the print editions of one of his spy series. Locke might argue this tactic was little different than that of authors who use the more traditional Kirkus reviews. The Kirkus Indie review service charges self-publishing authors $425 for a review: "If the review is negative, you will have the option of keeping it private and simply using the assessment as feedback to improve your craft. If it's positive, you will be able to use the review to market your book to consumers or to catch the attention of a literary agent or publisher."[32] However, this service follows a recent transfer of ownership as Kirkus struggles to remain viable in the new millennium. The *New York Times* reported on "The End of Kirkus Reviews" in 2009, and it was subsequently purchased by a new owner. Presumably, the self-publishing and review market is one way that the company hopes to stay afloat.

Sometimes review takers get taken themselves. Complaints of incomplete, shoddy, or tardy service can be found about Rutherford's Getting-BookReviews and Kirkus. On Craigslist, I found a disappointed reviewer warning others of a scammer who promised $60 per review of cool musical gear. But "when you want to get paid," you receive only $20 per review: "Stay away from this one, unless, you really enjoy writing boring product reviews for little compensation." Similarly, researchers who bought fake Twitter accounts found that out of their twenty-seven black-market merchants, eight attempted to sell them accounts already sold; one merchant tried to resell them the same thousand accounts three times.[33]

Although product reviews and ratings are obvious targets of manipulation, they are not the only things that can be gamed. In a study of reviews and ranking on a photography Website, I concluded that although it can be hard to quantify the qualitative, such as artistic merit, we attempt it nonetheless, and these quantitative mechanisms beget their own manipulation. At photo.net, people "mate" rated friends, "revenge" rated enemies, and inflated their own standing. Fixes to such manipulations often take the form of more elaborate quantification, such as labeling comments as "helpful," which are then manipulated as well. This insight is popularly referred to as Goodhart's or Campbell's law, concisely expressed by anthropologist Marilyn Strathern: "When a measure becomes a target it ceases to be a good measure."[34] For instance, online influence can be measured by Twitter followers and retweets, yet these (as well as YouTube and Facebook likes) are easily purchased by the thousands. In May 2012, the network and data security firm Barracuda set up three new Twitter accounts and was able to purchase between twenty and seventy thousand Twitter followers for each account, finding some twenty eBay sellers and fifty-eight Websites that sell fake followers. For a handful of dollars, hundreds or thousands of people will look at a Webpage for half a minute, click on the "share on Facebook" button, or retweet. Even a credit card stealing virus was modified to instead create false likes for the Instagram photo-sharing site.[35] All of the big sites have either purged suspect reviews or demoted them by changing their filtering algorithms to stay abreast of the fakes. For instance, in 2008, Amazon replaced its "classic" reviewer ranking system with a "new" one that demoted many of the top book reviewers. Amazon's intent was for the new system to favor more recent reviews that are rated as helpful while ignoring attempts to stuff "the ballot box."[36] This change caused Harriet Klausner to drop from her top seat to 2,057th place by April 2014. Conversely, Grady Harp, who has maintained a higher helpfulness rating, has fallen only to 47th place.

Clearly, this is a bit of a cat-and-mouse game, as demonstrated by those annoying puzzles that we sometimes must solve to post a comment.

CAPTCHA Cat and Mouse

The morning programming on my local NPR station is often "brought to you by" Angie's List: "Reviews you can trust." It also is supported by Reputation.com: "Working to protect individuals and businesses by ensuring accurate search results and authentic online reviews." This is an intriguing juxtaposition: at Angie's List, members pay a fee to be able to read and write reviews of local service providers, and at Reputation. com, clients pay to limit the damage from, among other things, negative reviews. These two companies (and consequently NPR) benefit from the comments at the bottom half of the Web. But what about the actual customers and merchants? I have discussed the value of reviews, the extent of manipulation, and some actual manipulators. But manipulation is not only the actions of mistaken, desperate, or immoral people. Comment manipulation is best understood in terms of an arms race in which people scramble to make their way on the Web.

This arms race is exemplified by CAPTCHAs, which became widespread around 2002 in response to the proliferation of free online services. (*CAPTCHA* is an acronym for "Completely Automated Public Turing test to tell Computers and Humans Apart.") Although the ability to comment from a Web-based email site or social network is a boon, such services are also abused to create sockpuppets and to post spam. Automated scripts (robots or *bots*) can easily create such accounts by the thousands. A CAPTCHA regulates the creation of accounts by requiring the completion of a task that is easy for most humans but difficult for computers. One prominent implementation, reCAPTCHA, uses garbled text from book-scanning projects. By solving a CAPTCHA, users prove that they are human and also contribute to the public good by helping digitize books.

Yet, "necessity is the mother of invention," and fakers respond with ever more sophisticated techniques to automatically solve CAPTCHAs. This then prompts the invention and deployment of new CAPTCHAs schemes. In 2010, researchers interested in "how good humans were at solving CAPTCHAs" had twenty-one popular schemes to test against. It turns out that they often are annoyingly difficult: when presented with the same challenge, three human subjects agreed with one another only 71 percent of the time.[37] Real humans are being left behind in the CAPTCHA arms race and many have simply stopped trying. When the video service Animoto replaced CAPTCHA with a different robot-frustration technique, completion of its sign-up form by humans increased by 33 percent.[38] Also, the efficacy of CAPTCHA has been undercut by the arrival of cheap human labor online.

Amazon's Mechanical Turk, which was launched in 2005, is named after the famous eighteenth-century chess-playing automaton. The chess-playing contraption appeared to be a machine but had a human chess master hidden inside. Similarly, "MTurk" appears to be an automated

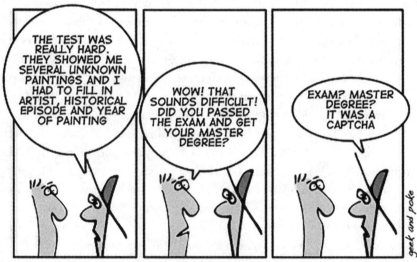

IN THE FUTURE SOPHISTICATED CAPTCHAS WILL LOCK OUT ANY BOT

Web service, but its tasks are actually completed by people. These *human intelligence tasks* (HITs) are things that people are good at, like categorizing images, translating text, and verifying information. Using MTurk, some people can request that other people perform HITs for them in a fast and automated way. Amazon noted that "Turk Workers" (or "Turkers") can work from home, on their own schedule, and "get paid for doing good work." On the other hand, "Requesters" have access to a "global, on-demand, 24x7 workforce"; they can "get thousands of HITs completed in minutes" and "pay only when you're satisfied with the results."[39] This system could also be used to solicit fake reviews and ratings, send spam, fraudulently click on ads, fill out forms, and create sockpuppets. In 2010, computer scientists at New York University measured how many new Requesters were submitting such HITs. Cleverly using Turkers to do the actual categorizations, they found that 41 percent of the 5,841 HITs were illicit.[40] Such fakery is now against MTurk's terms of service, and Amazon will seize the accounts of violators, though some complain of unfairness. One Turker faulted Amazon for not making these restrictions clear. He lost over $300, mostly gained from image categorization, when he responded to a HIT "like the one that asked me to type in a captcha image for 1 penny. That's right. One penny cost me more than $300 dollars."[41]

Although MTurk is no longer as palatable to manipulators, some sites traffic specifically in CAPTCHAs. Death by CAPTCHA (the "cheapest CAPTCHA bypass service") has plans that start "from an incredible low price of $1.39 for 1000 solved CAPTCHAs." Solutions must be offered before the challenge expires; this service claims that the average response time is fifteen seconds with an accuracy rate of 90 percent but warns that "our services should be used only for research projects and any illegal use of our services is strictly prohibited." Such illicit uses would include creating email accounts from which to post fake reviews and spam. Yet Death by CAPTCHA's page used to say "Don't let CAPTCHAs get in the way of your marketing goals! With Death by Captcha, you can bypass any CAPTCHA from any website. All you need to do is implement our API, pass us your CAPTCHAs and we'll return the text. It's that easy!" The ironies and hypocrisies of these services seem lost on their proprietors, such as the folks behind Bypass Captcha: they can be contacted only by a Web form that is protected by a CAPTCHA "anti-fraud code."[42] However, not all CAPTCHAs and accounts are valued equally. The authors of a 2013 study of "the underground market in Twitter spam and abuse" were able to purchase fake Twitter accounts from twenty-seven different merchants with a median cost of four cents per account. This is a relatively high cost compared to the fractions of a penny that Hotmail and Yahoo accounts go for because Twitter uses CAPTCHAs and phone-verified accounts. That is, a user is verified by receiving and entering a code that is sent to them on their phone. Sadly, on Craigslist one can find schemes that trick people into receiving such codes and handing them over to scammers. Other people who are duped by scammers are the unwitting downloaders of free pornography, movies, and software. For example, a scammer in need of CAPTCHA solutions can place the challenges on a pirated music site. The solutions provided by those downloading music are then passed back to the scammer to use. In response, more sophisticated CAPT-CHAs are being deployed, which themselves have commercial overtones. One CAPTCHA service requires users to watch a commercial and type in a product slogan to "prove one is human." Supposedly this delivers "1200% greater message recall than banner ads."[43]

This arms race, a competitive mix of exploitation and frustration, is characteristic of the larger dynamic of manipulation online. Increasingly, this arms race takes a commercial turn. Although we once solved CAPT-CHAs to help book-scanning projects, we are now forced to prove ourselves by interacting with advertisements.

"Pay to Play" and Extortion

Although I rely on online reviews, I know that they can be frustrating and duplicitous; the emergence of review sites and reputation services does not allay my concerns. Consider the following scenario. You wish to find a local service provider (such as a plumber or doctor) and wonder whether you should join Angie's List, which you have heard about on the radio, so that you can see what its members recommend. Before you pay the membership fee, you investigate what people say about Angie's List itself. After a simple search, you find warnings that it is difficult to terminate a membership, people are automatically resubscribed at a higher cost, and the fee would not necessarily keep a provider's family and friends from joining and posting inflated reviews. It also appears that service providers—including those not associated with the site—are pressured by aggressive sales people. One small business owner wrote that when Angie's List solicited him by noting he had a negative review, he was advised to "have all my good customers join and add their reviews. The customers would have to pay to join then pay to leave reviews. Pretty much like blackmail."[44] Is the desperation to have positive reviews the reason that some businesses give coupons to review writers?

How far is too far in encouraging positive reviews? At Amazon, a merchant of an inexpensive case for the Kindle Fire offered a refund to consumers who would "write a product review for the Amazon community." Although the merchant did not require five-star reviews, it hinted that they were appreciated: "We strive to earn 100 percent perfect 'FIVE-STAR' scores from you!"[45] If such reviews do not mention the refund, they are likely in violation of U.S. Federal Trade Commission (FTC) guidelines from 2009. The rules require the disclosure of nonobvious "material connections" (such as payments or free products) for consumer endorsements, including those "disseminated via a form of consumer-generated media." These guidelines were widely reported on but little noticed, understood, or enforced. In 2013, the FTC again attempted to remind the public of the requirements, even in "space-constrained ads" such as tweets.[46]

In addition to claims of "blackmail," one can also find accusations of "pay to play" and "extortion." Sites like Pissed Consumer and Ripoff Report are filled with such complaints about Angie's List, Yelp, and others. Yet many "consumer advocacy" sites are the subject of similar accusations: they are said to encourage false reports or refuse to remove reports unless they are paid or receive kickbacks from a reputation-management service. The worst pay-to-play schemes post "revenge porn" (embarrassing photographs or videos), mugshots, and photos of supposed sexual predators that can be removed for a fee. The now defunct site STD Carriers published claims about people's alleged sexually transmitted diseases. It refused to remove reports but would place a no-index tag on the reports of those who paid, rendering the report invisible to most search engines.[47]

From a legal perspective, Pissed Consumer and Ripoff Report are smart to refuse to remove reports. Section 230 of the U.S. Communication Decency Act (CDA) declares that "No provider or user of an interactive computer service shall be treated as the publisher or speaker of any information provided by another information content provider."[48] That is, a site that hosts information posted by users is not liable for that content. (Sites are obliged to remove illegal content, however, such as child pornography.) Some legal scholars feel that the immunity should not be extended to sites that knowingly host or solicit content that is defamatory or invasive of privacy. A judge felt similarly in a 2013 ruling against the gossip site The Dirty, although the decision is likely to be appealed. In any case, people are understandably alarmed to find negative comments about themselves online. Writer James Lasdun wrote of being harassed online, including in online reviews of his early work. Even if most people discounted the slander, he felt that "its real harm was the notice it gave that I was a person to whom such a thing could be done: that I had attracted an enemy who wanted the world to categorize me as an object of scorn."[49] While those who pay to have such content removed probably feel it is necessary, they also likely feel doubly abused.

The Racket of Reputation Management?

The main tactic of reputation services is to ensure that enough positive content is on the Web to drown out the bad. While customers hope to displace the negative content from at least the first page of search results, nothing is guaranteed. Sites that host negative comments also can be paid to lessen their complaints' impact. Pissed Consumer offers a "service" that allows clients to write a response or preemptively "get contact information of the consumer before [the] complaint is posted on the site." Ripoff Report offers a free rebuttal and a similar $90 per month program "that affords a business the opportunity to fix a problem with a complaining customer before the world has the opportunity to read the Report."[50] Other fee-based services can "make your search engine listings change from a negative to a positive" by changing the title of a report which is seen in search results. In 2010, Ripoff Report introduced a "VIP Arbitration Program," something that external "reputation management services can't deliver":

Is your company the subject of a false Ripoff Report? Do you feel a free rebuttal is just not enough? Would you like the opportunity to prove that the Ripoff Report about you or your business is false? Would you like false statements of fact redacted from the report? Rip-off Report has contracted with private arbitrators who have extensive experience, including judges, to decide your case. Clear your name once and for all the right way.[51]

The "filing fee" for this service is $2,000 to "pay for the arbitrator's time and for administration of the program."

Even widely known and used sites are suspect. Journalist Kathleen Richards wrote a series of "Extortion 2.0" articles that documented merchants' complaints that Yelp pressed them to advertise on the site. If merchants joined Yelp's sponsorship program, Yelp would feature a positive review on the merchant's page, cause the business to show up in related searches, and have the merchant's ads appear on competitors' pages but not vice versa. This, combined with two other facets of the site, led to allegations of extortion. First, the opaqueness of Yelp's filtering algorithms can be alarming: some reviews are shown by default and included in the overall star rating while others are shunted to the "filtered reviews" section. Yelp, like most all sites, keeps its algorithms proprietary, and this is a cause for concern and confusion across the Web. (Yet if Google's ranking or Yelp's filtering algorithms were known, they could be more easily gamed.) Second (and more troubling), Yelp employees write reviews. Merchants cry foul when they decline to advertise on Yelp and a negative review by a Yelp employee soon follows.[52]

Such complaints have yet to make much headway in U.S. courts. In 2011, two related legal suits against Yelp were dismissed. At issue was whether Yelp's filtering and aggregation of user comments stripped it of its CDA Section 230 immunity: that is, does Yelp's filtering of user-generated comment make it a publisher and consequently liable for the content? Regardless of Yelp's possible motives in the presentation of users' comments, the judge wrote that the plaintiffs did "not raise more than a mere possibility that Yelp has authored or manipulated content related to Plaintiffs in furtherance of an attempt to 'extort' advertising revenues." Cyberlaw scholar Eric Goldman opined that "This ruling makes clear that Yelp can manage its database of user reviews however it wants.... However, it doesn't mean that we as consumers will find Yelp trustworthy."[53] Other scholars contend that Section 230 immunity needs revision given the harassment, extortion, and defamation that occur within its shadow. Law professor Ann Bartow goes further, arguing that the "monetization of Internet harassment" creates "unsavory incentives" for the services to see the status quo (of harassment) perpetuated.[54]

Even the nonprofit Better Business Bureau (BBB), which has been an arbiter of reputable businesses since well before the Internet, has been tainted by monetization. In 2009, Bill Mitchell, the Los Angeles chapter's chief executive officer, was forced to resign because of allegations of a pay-to-play scheme. Mitchell, head of the BBB's largest national chapter, also originated the new letter-grade accreditation system, replacing the classic *satisfactory* or *unsatisfactory* labels with grades such as A plus and B minus. The scheme was so successful that it was adopted across the country. Yet after merchants were solicited by sales representatives (who earned 45 percent commissions on first-year memberships), they claimed that they were assigned unsatisfactory grades as a strong-arm sales tactic. The television news show *20/20* reported that unsatisfactory grades were elevated overnight after payment was received. To test this claim, producers were able to easily register a number of fictitious businesses, including an A plus "bogus firm named after Stormfront, a white supremacist group."[55]

Consumers are getting in on the racket as well. One reviewer claimed food poisoning, refused a $60 gift card at a restaurant of his or her choosing, and demanded $100 to refrain from posting a negative review. At Google's app store (that is, Android's Google Play), users often promise to raise a rating if a shortcoming is addressed: "Fix that and I'll give it 5 stars." For many years, app developers could not respond to reviews and would beg users to send questions, bug reports, and feature requests to them before posting a negative review. In 2013, when Google enabled developers to reply to reviews, the Android blog *Phandroid* quipped: "No longer will developers have to break down and cry when a user leaves a 1-star review for a feature they don't know how to use."[56] For those willing to pay $100 for a plastic card, the ReviewerCard is the "first-of-its-kind membership card and community for reviewers": members who display the ReviewerCard "enjoy premium service."[57] It is a fancy way of telling the restaurateur that cardholders might write a review. However, no restaurant is obliged to do anything because of the card, and a similar card is easily made for less than $100, so it is not clear who is exploiting whom.

All of this leads to conflict and headlines. Sometimes review sites are sued, such as Yelp for pay-to-play abuses and TripAdvisor for negligence in allowing defamatory reviews. More often, organizations try to make it impossible to be sued, or they sue their reviewers. In 2014, the food conglomerate General Mills updated its terms of service so that anyone interacting with it via social media waived the right to sue. Aggrieved parties that had *liked* Cheerios or any other GM brand would have to enter "forced arbitration" instead.[58] Medical practitioners are especially concerned about deleterious comments online. Reputation-management firm Medical Justice ("making healthcare safe for doctors") is infamous for its attempts to prevent bad reviews by providing its members with contracts that forbid patients from posting negative comments about a service provider. Another tactic is to prompt the patient (in fine print) to assign any copyrights in such comments to providers, who may then remove them at their discretion. In a bizarre case, a New York dentist sued a former patient for a review in which the patient claimed that the dentist overcharged him by thousands for what should have been a couple-hundred dollar procedure. However, Medical Justice refused to defend the terms of the contract (and has since abandoned the criticism clauses), and the dentist closed her office and was unreachable by her lawyers.[59] In an unrelated but similar incident, Utah consumers John and Jen Palmer claimed that their credit had been ruined by a company that wielded a nondisparagement clause. In 2008, the couple had expressed disappointment with KlearGear.com in a negative review on Ripoff Report. Although I have reservations about Ripoff Report, KlearGear's 2012 levy of a $3,500 fine for "nondisparagement" is also troubling. First, the clause was allegedly added to the site's terms of service after the Palmers' 2008 transaction. Second, when the Palmers refused to pay the fee, the nonpayment was referred to credit agencies, damaging the family's ability to get credit for essential purchases.[60]

Sometimes online arguments about reviews become real, as when an owner of a bookstore showed up at a surprised Yelp reviewer's door because he had said that her store was a "total mess." Also, news stories of crackdowns are common. In 2012, all of the following were reported. Amazon users noted that thousands of reviews had been deleted, although Amazon would not disclose numbers or a rationale. Facebook removed suspect likes, including 200,000 (about 3 percent) from a popular poker game page. YouTube purged more than two billion suspect views from channels belonging to large music labels. And of Yelp's thirty million reviews, roughly 20 percent were found to be filtered as suspect.[61] In addition to these purges, the big sites are also hoping to solve fakery via the "social graph," favoring the comments and activities of your acquaintances, although this will likely lead to new forms of manipulation.

Exploiting the Social Graph

On one hand, it's good to hear about services cracking down on fake comments, reviews, likes, and views. However, I suspect that this is as much for the benefit of the sites as it is for their users: external manipulators are in competition with the sites' own manipulation of their search results, ad placements, and "sponsored" reviews. Even if only true comments were posted at a site, the site would still seek to make money by manipulating the prominence of certain comments. There's a reason that *Consumer Reports* is widely trusted: it has no financial relationship with the makers of products, it even pays for all the products it tests. This is why I tend to be more trusting of Amazon among online retailers: they are relatively impartial among items sold and the reviews posted. This doesn't mean that Amazon is angelic: it has been accused of exploiting publishers and its employees to keep costs low, and it does sell ads on its sites for "sponsored products."[62] I also wish that Amazon did more to remove fraudulent reviews and helpfulness votes. Yet, while the particulars of its search algorithm and product placements are proprietary, Amazon makes a commission on sales, so it should be biased toward items that sell well—things that people actually want. Also, it's easy to see which reviewers actually purchased a product ("verified") and Amazon has formalized "kick-backs" by way of the Amazon Vine program: products given to reviewers are explicitly identified as such.

Even so, Vine reviews can be amusingly weak, such as one for the engineering book *System Identification: A Frequency Domain Approach;* a Vine reviewer wrote little more than "If your into math you will most likely enjoy this book."[63] And although more expensive (and complicated to purchase), illicit reviews can be had from those who actually buy the books from Amazon such that they are labeled as verified purchases. The persistence of such fakery is why the big Web services are now battling over a new paradigm of user comment that is known as the *social graph.* Instead of relying on strangers to inform consumption, people could use their friends' comments instead. People would no longer have to fear manipulation from strangers.

However, users still would not be free from manipulation by the services themselves. In spring 2012, Facebook demonstrated this with the deployment of Sponsored Stories—"messages coming from friends about them engaging with a Page, app or event that a business, organization or individual has paid to highlight so there's a better chance people see them."[64] (A Facebook "Page" is a paid-for account that is integrated into Facebook's advertising system with an unlimited number of friends/fans.) Sponsored Stories were labeled as such and could appear on the right side of the Facebook interface, but users were annoyed to find them also appearing in their News Feed (where Facebook limited their appearance to once a day). Those who had Page accounts also found that they were having a more difficult time reaching their fans. An average of only 15 percent of a Page's fans were "reached" and saw its messages. If those who had Page accounts wished to reach more fans, they needed to pay for Sponsored Stories. Cries of ransom, robbery, and extortion soon followed.

Worse yet, some users were surprised to find their photographs in Facebook ads. Facebook user Nick Bergus came across a beguilingly odd product on Amazon: a fifty-five-gallon drum of sexual lubricant that was listed for over a thousand dollars and had 146 customer reviews, although there was little evidence of any verified purchases. The reviews—and probably the product—were farcical. The highest-rated review was by actor and social media star George Takei (Mr. Sulu from *Star Trek*), who discussed how he tested the lubricant by spraying it on revelers during the San Diego Pride Parade. After Bergus posted a link on Facebook, commenting that the lube was "For Valentine's Day. And every day. For the rest of your life," his face began appearing in Facebook ads for the product. Similarly, Angel Fraley, an eventual plaintiff in a lawsuit against Facebook, appeared in ads for the Rosetta Stone language learning system because she had liked its online French course in hopes of getting a discount. She might not have received the discount, but she did appear in Rosetta Stone ads. Although this advertising practice was disclosed in the fine print of Facebook's terms of service, users challenged (and sued) that this was a consentless use of their identities, especially for users under the age of eighteen.[65]

Yet Sponsored Stories seemed to be effective for those who were willing to pay. One Web analytics company concluded that the ads cost roughly 20 percent less per click and per Facebook fan and were clicked on 46 percent more than standard Facebook ads. Although the service was soon bringing in $1 million in revenue per day, the bad publicity and the class action suit (in which Facebook agreed to pay $20 million) led Facebook to discontinue the service in 2013 as part of a larger restructuring of its advertising. However, Facebook's product marketing director Brian Boland stressed that "Sponsored Stories as an idea doesn't go away. Sponsored Stories as a product goes away."[66] Indeed, others have followed Facebook's model. Twitter has "promoted tweets" and the comment platform Disqus offers "Sponsored Comments," which "let businesses deliver a message to the people they need to reach" by pinning their content to the top of your discussion.[67] In 2013, Facebook's main competitor in the space, Google, announced "shared endorsements." While Google attempted to be more careful of upsetting its users, it still prompted some concern with the declaration that: "To help your friends and others find cool stuff online, your activity (such as reviews, +1s, follows, shares, etc.) may be used along with your name and photo in commercial or other promotional contexts."[68] And to help you generate sufficient comment, Google is also thinking about mechanisms that prompt users with "suggestions for personalized reactions or messages" based on what it knows about its users and their social networks. For instance, the patent application for this mechanism noted that "it may be very important to say 'congratulations' to a friend when that friend announces that she/he has gotten a new job," and if the busy user forgets to do so,[69] Google will suggest an appropriate comment. Facebook and Google want to help us generate grist for their mills.

Social networking sites might even benefit from external manipulators. Accusations of fraud surfaced again in 2014 when users noticed that paying Facebook to promote their content caused their engagement—and subsequent reach—to drop. *Reach* is the number of people who see content, and *engagement* is the number of people who click, like, or comment on it. At Facebook, users do not see all the posts of those they follow. Facebook parcels out users' posts to portions of their friends and fans based on how much engagement the content is getting and whether it was "promoted." People were concerned that although the promoted posts resulted in a lot of new fans, engagement numbers dropped even further. Online science communicator Derek Muller suspected that this was the result of *like* farms trying to hide their activity from detection. Manipulators often pay others (often in poor countries like Indonesia and Bangladesh) to like their content so that Facebook will share their posts with more users. (Facebook sees this as engagement, infers that it is useful, and spreads it further.) To avoid detection, these scammers hide their fraudulent activity in a storm of random likes, often landing on recently promoted posts. Although Facebook could boast about the benefits of advertising based on the number of new fans, people were falling further down a hole. Promoted content on Facebook is clicked on by the farms (to mask their fraudulent clicks), which subsequently drives down relative engagement because real fans are outnumbered by the now dormant fakers. Muller complained that Facebook was doing nothing to address the problem and that "they have since benefited from those 80,000 likes" because "I paid to boost posts out to these useless likes." Others concluded that the hundreds of thousands of dollars that they paid to Facebook had been wasted.[70]

The dynamics of all of this is captured in a Geek&Poke Webcomic in which two pigs discuss the benefits of free services like Facebook and Google. One pig says, "Isn't it great? We have to pay nothing for the barn," and the other responds, "Yeah! And even the food is free."[71]

A "Loss of Innocence"

"E.Z." (a pseudonym) is an Amazon Top 500 Reviewer who has posted over 250 reviews (a relatively low number for top reviewers), and he has accumulated nearly two thousand "helpful" votes on those reviews. He started using the Internet in the 1980s and has shifted from being an active and public participant on early Internet forums to tending to his own online activities and Websites. He told me, "I am not a twenty-year-old who is always posting on his Facebook account and telling everyone what he ate for breakfast every day." Although many Amazon Top Reviewers focus on items that are easy to review, E.Z. believes that quality matters more than quantity, so he is selective in what he reviews and is often exhaustive in scope.

I first encountered E.Z. while I was looking for a digital voice recorder to use for interviews for this book. My purchase was influenced by the depth of his seven thousand–word recommendation. It was so large that it exceeded Amazon's word limit and continued in his review's first comment. Over a hundred other appreciative comments followed. When E.Z. began reviewing he started slowly and did it as "a cathartic way to vent about my frustrations using bad products and to also warn other people not to buy those products." Since then, E.Z.'s reviews have been mostly

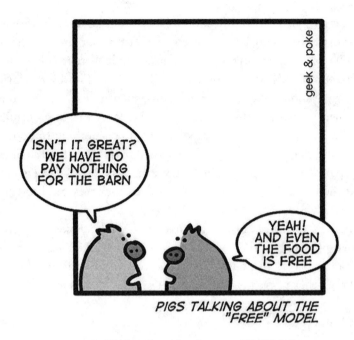

PIGS TALKING ABOUT THE "FREE" MODEL

positive, which he attributes to his thorough research and comparison shopping. Given his long history with online comments, I was interested in E.Z.'s sense of how things have changed over the years, especially his thoughts about Amazon's Vine program in which top reviewers are given free products to review.

E.Z. told me that Amazon invited him to participate in Vine in the spring of 2008 and that the several thousands of dollars' worth of free merchandise every year does motivate him to write more reviews. (To stay in the program, reviewers originally had to review 75% of the items they received; this was later moved to 80%. In 2013 some reviewers objected to a proposal that they must review 100% of received products within thirty days.)[72] His time in the program "has filled every room in my house with free goodies," including books, four digital cameras, a large-format photo printer, a $1,400 home security camera system, other electronics, and lots of software. Other Vine reviewers have received refrigerators, washing machines, and dryers. He conceded that this likely prompts "the three amigos of greed, jealousy, and envy" as reviewers "clamor for higher reviewer rankings and better products than their fellow Vine members."

An aphorism among those who study human motivation is that extrinsic motivations crowd out intrinsic ones. This means that an external imposition (such as a payment or punishment) often lessens people's internal motives related to feelings such as autonomy, mastery, usefulness, and self-esteem. A classic example of this is a daycare center that struggled with some parents who frequently were late in picking up their children. When researchers experimented by adding a fine for tardy parents, the unwanted behavior actually increased. The late parents saw the "fine as a price" that they were willing to pay. The external punishment displaced their internal motive to see themselves as responsible parents who abided by an implicit social contract with the teachers. Even when the fine was removed and the researchers left, the high rate of tardiness continued. At Amazon, top reviewers report that they are motivated by intrinsic values such as self-expression and enjoyment.[73] However, Vine might be changing that. E.Z. noted that if he no longer received free Vine products, "my interest in writing online reviews would likely plummet, and I would not care how low my Amazon Reviewer Ranking would plummet because of that." His motive to warn others away from a bad product might remain, but he rarely purchases a dud now and probably would not bother to write positive reviews.

Finally, E.Z. felt that there has been a "'loss of innocence' with online discussions, comments, and reviews of products." Before 1996, when advertisements on the Internet were verboten, E.Z. enjoyed authentic discussions, such as those about bicycling and mountain bikes:

Back then, there was not a single advertisement posted on those newsgroups and no mountain bike manufacturers were posting comments or (as some manufacturers do now with Amazon reviews) posting fake comments that raved about a specific brand of mountain bike. So if someone was interested in buying a mountain bike during those Usenet days, they could either post questions themselves or read other postings on a newsgroup and all of the raves, criticisms, and good or bad comments about products were much more sincere and authentic than what we have to sift through on the Web these days.

So with Amazon's Vine program, with Amazon's reviews in general, and with reviews on other sites like Yelp, TripAdvisor, etc., I feel that there has been a "loss of innocence," and this loss of innocence is often happening simply because money is involved, either with the free products being given by Amazon to their Vine reviewers or with manufacturers realizing the increasing importance of user reviews/comments in this globally socially networked world that we now live in.[74]

As seen in this chapter, many are complicit in this loss. People traffic in the illicit markets of comment, some click a *like* in hopes of a discount, diners loudly discuss Yelp when their server is nearby, restaurateurs give coupons in exchange for reviews, authors ask friends to write reviews (or do so themselves), sockpuppets edit Wikipedia biographies, pundits purchase fake followers, and sites profit by manipulating users' praise and pillory. Much of this behavior is driven by the high value of comment today, an obsessive desire to rate and rank everything, the dynamic of competition, and the sense that everyone else is already doing it. The world of online comment is quite different from that of early likers, like E.Z., who reviewed for the love of it. At the bottom of the Web we are increasingly tempted to become manipulators, and as we do so we lose something in the process.

4

Improved: "Be More Constructive with Your Feedback, Please"

Other rappers dis me; say my rhymes are sissy.
Why? Why exactly? Why? Be more constructive with your feedback, please.
Is it because I rap about reality? Like me and my grandma drinking a cup of tea?
There ain't no party like my nanna's tea party. Hey! Ho!
—Flight of the Conchords, "Hiphopopotamus vs. Rhymenoceros"

The fear of public speaking is supposedly so great that comedian Jerry Seinfeld once quipped that people at a funeral would "be better off in the casket than giving the eulogy." Is the basis for this joke true? Richard Garber, the blogger behind *Joyfully Public Speaking*, investigated and found that on a 1993 survey that asked people about their fears, 30 percent of those surveyed said they feared death, and 45 percent reported a fear of public speaking. However, although more people did report a fear of public speaking, they never were asked to compare it to death.[1] Yet this quibble is easily forgotten in light of the keen anxiety many feel when they speak in public. They are afraid of disparaging comments that might arise in others' thoughts or, worse yet, comments that might be furtively exchanged between their peers. In fact, professionals make a living studying "communication apprehension" and treating "glossophobia," from the Greek *glōssa* (tongue) and *phobos* (fear).

One way that people improve their public speaking and lessen their anxiety is to practice in a supportive space. The Toastmasters International organization aims to provide an environment where its members can improve their speaking and leadership skills. At a Toastmasters meeting, participants encounter guidebooks, roles (such as the "time keeper" and "ah counter"), evaluations, and competitions—all of which are intended to move people toward "competent communication." The irony of Toastmasters's focus on evaluation and competition is that a source of anxiety is a fear of judgment (what are others thinking or saying about me?). And in the age of the Web, such judgments are seemingly more frequent and public. For instance, the online commentary about someone's presentation is known as the *backchannel*. Recently, academic tweeter Kathleen Fitzpatrick wrote that she was feeling frayed by quick and unthinking comments on the backchannel: "I've done this, probably more times than I want to admit, without even thinking about it. But I've also been on the receiving end of this kind of public insult a few times, and I'm here to tell you, it sucks." The problem is that in addition to the "the real generosity, the great sense of humor, the support, the engagement, [and] the liveliness" that Twitter engenders, it also "produces a kind of critique that veers toward the snippy, the rude, the ad hominem."[2]

Although evaluations at Toastmasters meetings typically are short (often three to five minutes), the intention is to provide useful comments to the speaker. The evaluator should avoid the "three Cs" (criticize, complain, and condemn) and instead perform the "three Rs" (review, reward, and recommend). Evaluation should *h*elp the speaker improve, *e*ncourage another speech, *l*ift self-esteem, and *p*rovide useful recommendations (i.e., H.E.L.P.). Good feedback is so important at Toastmasters that evaluation is evaluated, which can also be a source of anxiety: as one club member wrote, "I am scheduled to evaluate a brilliant speaker tomorrow at our toastmasters meeting. Needless to say I've been feeling a bit apprehensive."[3] But the group provides guidelines, tips, and many opportunities to practice. It even hosts competitions to determine who can give the most masterful evaluation—examples of which can be viewed online.

Toastmasters's culture shows that giving and receiving feedback can be a rewarding and difficult practice, much like public speaking itself. This can be even more challenging in the online realm. As seen in Fitzpatrick's concern about the backchannel, the shortness of comment can lead to slights. The asynchronicity of comment (its disconnect from the bounds of immediate interaction) can cause context to be lost (or "collapsed"). The potential for both constructive critique and "snippy" tweets is heightened, as is the emotional work required of users to manage it all. In this chapter, I address *feedback*: comment that is intended (or at least expected) to be seen by the person it is about, its *object*. To do this, I look to three online comment cultures: peer feedback in an online course, in a less formal writing community, and in communities where the line between feedback and collaboration blurs. Among the many types of comment, what is most salient about feedback is that it can be personal. It is an expressed reaction to something that often is close to the core of another's self. Consequently, I will show how posting and reading comments on the bottom of the Web can require a brave heart and "strong ears."

Bridging the Gap with Feedback

According to the *Oxford English Dictionary* (*OED*), the word *feedback* is a creature of the twentieth century that was first used by electrical and mechanical engineers and later applied to biological systems. Norbert Wiener, the famed proponent of cybernetics, spoke of communication feedback within social systems in the 1950s. This sense of the term then appeared within the *OED* in a 1959 quotation that speaks of a person who adjusts a speech based on the "'feed-back' from the listeners." In 1971, it was used in the sense of this chapter: "We began to get a fairly good feedback from most people."[4] This reflects that people sometimes speak of *positive* (good) and *negative* (bad) feedback, distinguishing between what works well and what can be improved.

The emergence of the term *feedback* also paralleled a new perspective in teaching—and a few more new terms. In 1967, Michael Scriven coined the terms *summative evaluation*, which assessed the performance of the learner, and *formative evaluation*, which assessed the effectiveness of the curriculum. Yet, almost two decades later, his colleague, Royce Sadler, lamented a continued emphasis on test scores and statistics: "Only cursory attention has usually been given to feedback and formative assessment." Sadler advocated for a continued shift in evaluation toward how teachers can "shape and improve the student's competence."[5] This is now the dominant sense of the word among teachers.

Today the idea of formative evaluation is that students have a level of competence that is short of a learning goal and they can be helped to bridge the distance by feedback. Students can be instructed on how to traverse the gap, but a more substantive type of learning occurs when they are able to specify their own learning goals, assess their own competence, and devise techniques for bridging the gap. Facilitating such *metacognition* in the learner is not easy. For example, I assess my students' assignments on the basis of a rubric, and in the hopes of advancing deeper learning, I have asked students to give feedback to their peers using the same rubric: does the assignment substantively engage the appropriate concepts, does it show an understanding of those concepts, is it well written, and does it conform to typical scholarly conventions? Not only did students fail to use the rubric in the execution of their own essays, but the feedback that they sometimes gave their peers focused on fixing a couple of typos and pronouncing the essay worthy of an A. One issue is that students generally fear "breaking rank" or "throwing someone under the bus" when it comes to peer evaluation. Furthermore, in the language of educational psychologists, the students' "unconscious incompetence" (not knowing what they don't know) and my "unconscious competence" (being blind to the scaffolding needed by the students) failed to intersect. Good teachers compensate for their expert blindness and help students move toward "conscious competence." Unfortunately, my own mistake of not teaching students how to give good feedback has been replicated on a larger scale.

A *MOOC* is a massive and open online course, and at the 2013 World Economic Forum in Switzerland, MOOCs were center stage in the panel about reinventing higher learning. (The term parallels *MMOG*, or massive multiplayer online game, such as *World of Warcraft*.) The strongest advocates of MOOCs tend to be from recent online ventures such as Coursera and Udacity. However, even traditional educational institutions are argued to benefit from them because they can enroll students from outside their geographical reach in new degree or certificate programs and they can lower their costs by shifting some traditional courses online. The stories about far-flung students who take classes at MIT and Harvard are compelling, but MOOCs are criticized for their uneven quality, low completion rates, and cost savings that are suspected to come at the expense of teachers' jobs and students' learning.

Beyond cost savings, some claim that the MOOC educational experience is superior to in-class learning. Coursera, a commercial MOOC, notes that its online platform provides an opportunity to exploit mastery learning and peer assessment. In mastery learning, students can retake randomized versions of assignments until they know the material rather than following the instructor to the next topic while they still are unprepared. Also, instructor feedback traditionally is "often given weeks after the concept was taught, by which point the student barely remembers the material and rarely goes back to review the concepts to understand them better." Automated and instantaneous assessment can offer significant benefits. Yet, computers are not effective at assessing complex assignments like poetry and business plans, so students' metacognition can be advanced by recognizing that "the best way to learn is to teach." Coursera claims that it does two things: trains students to use a grading rubric to give accurate feedback to their peers, and uses crowd-sourcing to "take many ratings (of varying degrees of reliability) and combine them to obtain a highly accurate score."[6]

Laura Gibbs was one of the five thousand students eager to begin Coursera's fantasy and science fiction course. She also is an instructor at the University of Oklahoma, where she has taught online courses for over a decade. She focuses on myth and folklore and prefers online teaching because it allows her "to be a far better teacher, and I also think that it brings out the best in the students too."[7] Despite being an expert on the topic and experienced with online teaching, Gibbs took the course because she loved the reading list and was curious about how such a large course would work. On her blog and as later reported in a story at *Inside Higher Ed*, she reflected on the challenges of the course, especially peer evaluation.[8] During the course, students wrote short essays of 270 to 320 words and evaluated the essays of four peers. Evaluations consisted of a score of 1, 2, or 3 and at least thirty words of feedback.

Gibbs noted that some students were motivated to give helpful evaluation or were motivated to continue because of it but that many left the class because of poor and abusive evaluation. There was little instruction on how to give useful feedback, and she wondered about the lack of response from Coursera staff members to questions about abuse on the class discussion board. She also found that "there's a discussion board thread which consists solely of making fun of other people in the class. Everyone is a potential subject for abuse; anonymous is an equal opportunity scoffer."[9] In addition to personal insults, there also were plenty of one-word evaluations, like "ug," "no," and "terrible." What she found "most bizarre and strangely cruel" was a comment that spelled out the words "one" through "thirty":

this is because our comments are supposed to be 30 words long. The software does not police this (hence the abundant one-word and two-word comments: "good!" or "liked it!")—but the idea that someone would deliberately put in a comment like this to meet the word count shows that there are some serious problems with the feedback culture in the class.[10]

Complicating much of the experience was the varied skill levels of the students, which in part stemmed from the Web's global reach. Some wrote at a remedial level, some wrote English as a second language, and some contributed in foreign languages (or as translated by online tools). One person was accused of making words up by those with smaller vocabularies. Not surprising to anyone who has followed the English Wikipedia, there were heated arguments about American versus British spellings.

Coursera is a new enterprise and perhaps will figure out how to facilitate a robust learning community with a strong comment culture. This is no easy task. Even in traditional classrooms, research indicates that team-based learning and peer feedback are mixed bags that work only when the right levels of motivation, training, and expectation are present. Even then, results vary: well prepared and motivated students may thrive, but others fare poorly.[11] Interestingly, there is already an online community of learners who are interested in fiction and fantasy and have developed a robust culture of making constructive comments.

Beta Reading and Constructive Criticism

The *fan fiction* community is a sprawling cultural phenomenon in which fans write their own fiction within or at the intersection of official *canon* universes. (In the *slash* subgenre, for example, there are stories in which *Star Trek*'s Captain Kirk and Mr. Spock are lovers.) These amateurs have built a remarkable culture in the truest sense of the word *amateur*: they do it for the love. And because of their appreciation of the stories and characters within their particular fandom, they want to help others improve the stories they tell through *beta reading*—the giving of feedback on others' writing. The term is borrowed from computing culture, where *beta* software is typically complete but untested. *Beta testers* are willing to use a software program, encounter bugs, and submit reports or fixes (called *patches*). Similarly, *beta readers* read the initial rough drafts of other people's stories. Much like the participants in the Coursera class, *fanfic* authors' expectations and skills vary. After many years and endless debates, however, they have developed a novel lexicon and culture of feedback.

In the summer of 2008, "synecdochic," a fanfic writer, posted a Live-Journal entry entitled "'Cult of Nice' versus 'Cult of Mean,' Round 2847, Fight!" synecdochic is an online pseudonym of Denise Paolucci, a prominent figure in online journaling sites such as LiveJournal and Dreamwidth. Both of these services are similar to blogs, but journaling actually preceded the popularity of blogging—and never seemed to have gotten as much respect. In her post, synecdochic noted that the long-debated issue about the tone of comment had arisen again and was partly the result of confusion about different types of comment. In synecdochic's understanding, *concrit* (constructive criticism) is "editorial feedback" that is offered to the author "with specific suggestions on how to improve." *Commentary* is a reflection of the reader's response that is intended for the author but does not necessarily include specific suggestions for improvement. *Review* is like commentary but is intended for others, and *recommendations* are persuasively positive. She defined *flaming* as "hurtful and non-productive commentary" that is intended to cause emotional injury to the author. And she wrote that the word *critical* can be problematic because some people perceive it as necessarily negative and others do not.[12] With a vocabulary in hand, she enumerated a collection of additional "propositions that I operate under" when it comes to feedback, including that people want different types of feedback and have differing tolerances for the type and tone of comment. More interestingly, her propositions focused on the notion of space.

In the digital age, the scope and scale of comment have changed. In the past, feedback's scope was relatively local: in a public speaking group, candid feedback was shared with the speaker or at least remained within the confines of the room. Online, unsolicited comment can easily find its way to you and everyone else. As scholar Helen Nissenbaum writes, we develop social norms in particular contexts and "information technologies alarm us when they flout these informational norms—when, in the words of the framework, they violate contextual integrity." Similarly, danah boyd describes this as "context collapse," meaning that the lack of "spatial, social, and temporal boundaries makes it difficult to maintain distinct social contexts."[13] Comment's reactivity, shortness, and asynchronicity mean that it is especially contextual but that its context also is easily lost as it is forwarded and retweeted. Just as *hypertextual* means "beyond" textual, or existing within a web of links, I use the term *hypotextual* to indicate how comment's links to context are easily severed. The implication of this for feedback is that most people make judgments and perhaps gossip about others, but such gossip will only spread so far. Online, gossip is only a click away. (In chapter 7, I return to how hypotextual comment leads to confusion and controversy.)

My appreciation for beta reading and concrit does not imply there is never negativity: *snarky* (cutting or snide) comment is its own genre within fanfic, and it is associated with the 1990s television show *Mystery Science Theater 3000*. In a typical episode of *MST*, a man and his robot sidekicks watch bad movies while riffing on awkward writing, bad acting, and low-budget effects. Their silhouettes appear seated below the screen as they talk to one another or throw popcorn, and it feels like watching a movie accompanied by the jests of your wittiest, pop-culture-addicted friends. In fanfic, this has inspired a practice known as "MSTing" or "sporking" (a spork is a combination of a spoon and fork). The Website *Urban Dictionary* defines sporking as "a line by line critical analysis of fanfiction, typically of the Utterly Horrible or occasionally So Bad It's Good variety. Derived from the term 'Sporking one's eyes out,' implying that the fic is so bad that most people would prefer to attack their own eyes with sporks rather than read it."[14] (The practice of reading content that that is thought to be awful or objectionable is known as *hate reading*.)

Members of the fanfic community negotiate how and when sporking occurs. In her journal entry "'Cult of Nice' versus 'Cult of Mean,'" synecdochic posits that although authors can request others not to comment on their stories and commentators need not avoid "discussing a work" or "only offer unstintingly positive reactions," they should be aware of the virtual spaces where they do so. In "one's own space" (a mailing list, blog, forum, or wiki), writers and readers can express their requests or opinions and expect others to abide by those requests, even if they cannot control what happens elsewhere.[15] Even if people do as they please in their own spaces, the Web offers only a tenuous veil of protection. "Egosurfing" (searching the Web for one's name) or reading pages that link to your work can result in a rude disappointment. As one sporking site declares, the first rule of sporking is "don't link any of these sporking's back to the authors in question."[16]

Of course, the complaint of meanness is perennial in both mainstream literature and fanfic. Sarah Fay, writing online for the *The Atlantic* in

2012, chronicled a history of complaints between those who find book commentary to be too soft and those who find it too mean, including writer's Zadie Smith's 2001 plea that book critics be "more human" and try to impart "some kind of useful advice."[17] David Denby, in his 2009 book *Snark*, wrote that contemporary snark is "ruining our conversations." Unlike irony and satire, which can be used in the public service, snark has zero civic interest: "Snark is hating on the page. It prides itself on wit, but it's closer to a leg stuck out in a school corridor that sends some kid flying." It pretends to be in fun, which has the "doubly aggressive effect of putting the victim on the defensive" because "no one wants to argue with a joke."[18] Even within the sporking community, there are lamentations that the good old days have given way to "flaming and more personal vendettas" and that the love sporkers had for flawed material (because it "told you about yourself, or the author, or it was a guilty pleasure") has faded.[19]

Beta Bruising and the "Feedback Sandwich"

Even when feedback is well intentioned and welcome, it is not necessarily easy to give or receive. When giving feedback, it is easy to focus on the person rather than the work, to make overly broad generalizations, to complain without offering a path forward, to presume that everyone else feels as you do, and to presume to speak on everybody's behalf. When receiving feedback, writers find it easy to become upset or defensive or to ignore what could be helpful. Hence, many of the reflections on beta reading provide tips on the process; some even recognize that one is building a relationship in which the author and reader have responsibilities towards one another. Miriam Heddy, a fanfic author, wrote that "I've had two major, long-term betas in my fanfic writing life. And in both cases, the betas beta me because they like my writing enough to want to see it get better." In time, both writer and beta commenter come to know each other's idiosyncrasies and may become friends. For the relationship to remain constructive (rather than simply validating), however, a balance of honesty and support must be struck. If both parties find themselves "spending more time patting each other on the back than mocking each others' idiosyncrasies, you may be in the wrong relationship."[20]

Beta reading guides commonly recommend that collaborators be explicit about their needs and expectations. If writers want feedback on what they did well or what they should improve, they should say so. Even so, beta readers should respect that the story belongs to the author and should not coerce the premise of the story or characters to their own liking. Additionally, a *Star Wars*–related site notes that communication should be explicit and diplomatic. One tip toward that end is to use the *sandwich technique*: "Praise, then critique, then praise again. Don't critique the author, critique the story." Be specific, and offer a suggestion. Instead of commenting that "Yoda sounds out of character," explain, and offer a solution:

Yoda doesn't answer questions in a straightforward way. Yoda would take his time, and he's more likely to lead the questioner to the answer by asking questions of his own. The word "me" doesn't sound right. How about changing that line to something more like: "Think you do, that immobile am I? Mace Windu from the Council, is the only other."

I have read at least a dozen guides on feedback in fanfiction, and, more broadly, there are millions (according to Google) of pages that mention feedback. Most of the popular guides mirror the findings of researchers. For instance, "destructive criticism" (criticism that is harsh in tone, nonspecific in nature, and focused on internal causes) has been found to undermine people's sense of confidence and self-efficacy, reduce people's self-set goals, and impair their performance.[21] In terms of negotiating the type of feedback desired, researchers have found a relationship between the character of feedback and the expertise of the recipient. Novices respond more to positive feedback ("this is what you did well") because it reinforces their commitment to a goal. Experts respond better to negative feedback ("this is what you can improve") because it helps them judge their progress toward a goal.[22] Valerie Shute, a researcher in educational psychology, conducted a literature review of one hundred scholarly documents on *formative feedback*. She concluded that feedback providers should focus on the task, not the learner; present specific and clear feedback in manageable units; give feedback in writing rather than in conversation; and for difficult tasks, give immediate feedback to help the student along but delay feedback on simple tasks so as not to interrupt.[23] Finally, not all "feedback sandwiches" are equally effective. Communication scholar Clifford Nass has noted that some sandwiches suffer from *retroactive interference*, which occurs when negative remarks cause us to forget what was just said because we are thinking so hard about the criticism: "the criticism blasts the first list of positive comments out of listeners' memory," and the concluding "positive remarks are too general to be remembered." Instead, feedback providers ought to offer broad praise, provide brief criticism that is focused on specific steps toward improvement, and conclude with lengthy and detailed positive remarks.[24]

Because even the most skillfully delivered feedback can be bruising and a lot of feedback is less than skillful, the exchange of feedback can be seen as a type of *emotion work*. This notion was introduced by the sociologist Arlie Russell Hochschild in her 1983 book *The Managed Heart: Commercialization of Human Feeling* to describe the "management of feeling" and its manifestation in interactions. In Hochschild's study of flight attendants, the ability to smile, regardless of the unreasonableness of passengers, was as important as any other aspect of their job.[25] So in exchanging feedback, each party works to make the experience palatable and productive. For instance, writers can distract themselves from their own misfortunes by browsing through the pages of *Pushcart's Complete Rotten Reviews and Rejections: A History of Insult, a Solace to Writers*. Reading others' rejections reminds writers that they are not alone and that even the best writers are rejected. As the editors of the collection wrote, "One of the pleasures of this wicked collection is watching the great being terribly wrong about the great." The collection also shows that even the greats—those presumed to be supremely confident—can be vulnerable. The science fiction master Isaac Asimov wrote to the editors of *Pushcart* that he followed the advice of a colleague to stop reading a review at the first unfavorable adjective: "I have done that faithfully and, as a result, I have no bad reviews to send to you. I also throw away good reviews, by the way, but I read them first."[26] Similarly, disappointed researchers can console themselves with a study that found that journal articles that were initially rejected received more citations on average than those that were accepted when first submitted.[27]

Even the anticipation of solicited and constructive feedback can make a writer uncomfortable, as Joyce Shor Johnson, a young adult novelist, notes: "let's say you have a beta reader and they've read your manuscript. Waiting is hard. Every now and then you wonder if they hate it." After the feedback arrives, Johnson recommends taking a deep breath, grabbing a glass of wine and highlighter, sitting back, and reading the feedback: "If you find something that makes you want to stop reading, highlight it and continue reading. Read it all the way through and don't think defensive thoughts." After the initial reading is done, put it aside for at least twenty-four hours. Then begin with the easy stuff, grammar and punctuation, before moving on to the more difficult issues that you have highlighted (those you can parcel out as your emotional reserves permit).[28]

And what about evaluations that are "just plain wrong," as physics professor Heather Whitney asked about students' comments: "Even when feedback is hurtful and/or inaccurate, can we still use it to improve our teaching?"[29] Most of the ensuing comments on her post reflected the sentiment that end-of-semester evaluations are of little use and that it is more useful to ask for feedback while the course is in progress. Whitney's question is relevant to more than just teachers. Among computer programmers, feedback can be especially harsh, which leads some to agree with Friedrich Nietzsche's line "what does not kill me makes me stronger." Software developer Chris McDonough writes that because complaining "appears ingrained in human nature, and is never going to change," people should adopt an attitude of using all comment, no matter how negative, to improve. In a blog entry entitled "In Praise of Complaining, Even When the Complaints Are Absurd," McDonough notes that when others opt to "complain cluelessly [and] bitterly in public," he still uses it to his own advantage. For example, the documentation he maintains about one of his projects "is almost pure spite-driven development. Things are only added to it, under duress, spitefully; it's largely a direct reaction to misinformed complaints." Yet, such a document is useful because he is then forced to explain the motivation behind his design decisions to avoid additional misinformed complaints.[30] After he transmutes the clueless and mean-spirited comments into something useful, McDonough is also likely to "flip the bozo bit" and stop extending the benefit of the doubt to the complainer. Avoidance, commiseration, triage, and even transmutation are all strategies that we have developed to deal with feedback.

Tweak Critique: "Good Form" or Unwelcome?

Photo.net is one of the Web's oldest photography sites. Its discussion boards can be traced back to 1993, and it describes itself as a site for serious photographers to learn by sharing and critiquing one another's photos. In March 2005, Patricia Minicucci posted her fourth message to the site's discussion board asking about feedback etiquette:

I am a relatively recent member of Photo.net and I actually enjoy contributing critiques (versus ratings), mostly because I learn a lot by looking at the work of others.... Initially, I would never have presumed to offer an alternative view of an image. Then, a member here offered me a revised crop on one of my own photographs that improved the image significantly. Frankly, I was grateful for the input and realized that the tweaked version spoke more eloquently than words ever would have. So, thereafter, I began to post some tweaked versions (usually just cropping/alignment/color balance stuff).[31]

Not everyone appreciates these tweaks, though, and one member responded, "You guys just don't get it. I for one do NOT appreciate you taking my photograph and doing what you feel is the right thing. Please don't ever do that again to my photography." As has been noted, feedback is not easy to give or receive, and communities of learners must negotiate their expectations and feelings about the practice. As Minicucci asked, "Is it generally considered good or bad form to offer a tweaked version of an image submitted for critique?"

At photo.net, some view *tweak critique* as the most important feature of the site, and it is explicitly permitted in the site's terms of service. In the comments that follow Minicucci's question, some members noted that a "picture is worth a thousand words" and that tweak critique is like book editing: "No writer should ever resent a good editing." Some requested that others refrain from making alterations to their photos. Many were happy to respect such requests because there is little to gain by giving someone feedback in a way they have explicitly declined. Additionally, one member noted that tweakers should place a water-mark or banner on photos that they alter so that they are not confused with the original.[32]

The ease, value, and acceptance of tweak critique in this community are novel, but as reflected in the comment about "good editing," the practice has precedents. For example, a photographer might put an assistant in charge of lighting but tweak and explain the setup to further the assistant's learning. However, this is a relatively intimate relationship. At photo.net, a stranger might manipulate a portrait of the photographer's child. Tweak critique has the virtue of being a specific, concrete suggestion, but such specificity can sometimes be unwelcome.

Tweak critique has become salient now because technology finally makes it easy. This notion is often spoken of as an *affordance,* a term that was coined by visual psychologist James Gibson and popularized by designer Donald Norman in the book *The Design of Everyday Things.*[33] Norman was concerned with "perceived affordance" or the extent to which a user is able to perceive that some action is possible with an object. His complaint was that many everyday objects are designed poorly, such as a pull handle on a door that can only be pushed open. Digital imaging significantly changes the affordances that are available to photographers and their critics. And the affordances of digital, networked communication affect more than photographers.

"Be Bold," but Prepared to Be Flamed?

Tweak critique can be found many places online. At Wikipedia, for example, a maxim tells readers to "be bold": "If you see something that can be improved, improve it!" A change can be suggested on an article's talk page, but Wikipedia wants people to "just do it" (without being reckless). Although editing others requires "some amount of politeness," "Wikipedia not only allows you to add, revise, and edit articles: it wants you to do it."[34]

The process that gives Wikipedians the ability to be bold is known as *version control*: a system of keeping, managing and merging revisions of a work. Most any change can always be reverted, restoring an earlier version with little or no harm done. Version control also helps projects integrate improvements. In software development, a programmer might receive a bug report that documents a confounding problem or desired feature. With the increasing popularity of free and open-source software, which lets others modify a digital work, people might even send a *patch* that fixes the bug. That is, people make a copy of a work, make their improvements, and then extract the differences (or *diffs*) between the two versions and sends those improvements to the original developer. This type of comment and tweak critique is in the textual rather than visual domain. Developers find it more efficient to receive a patch rather than a bug report—even if the report is well specified, which most are not.

As evidence that bug reports and patches are a type of feedback, they too can generate plenty of arguments. Much like writers of prose, software developers argue about bug reports and patches that are thought to be disrespectful to or divergent from the original developers' intentions. Developers complain that bug reports are ambiguous or confusing, and those who report them can be offended when their reports are labeled "INVALID" or "WON'T FIX." Patches can be rejected for being sloppy or "brain dead." The feedback culture around the Linux kernel (which mimics the confrontational style of its creator, Linus Torvalds) is famously harsh. Linux is a free source software project in which development and scolding happen in the open. Greg Kroah-Hartman, a Linux kernel maintainer, displayed this attitude when he complained about a patch:

And people wonder why kernel maintainers are grumpy.... This patch is why.... The programmer thinks they are smarter than the kernel and have silenced the nasty messages it was spitting out at them.... To quote the old IBM phrase, "THINK." To which I'll add "or you will be mocked."[35]

Most would agree that disabling diagnostic errors instead of fixing the errors you are causing is egregious. The culture of Linux development is to police such behavior with mockery and insult. In fact, the Linux operating system project was born during a famous feud between Torvalds, then a computer science undergraduate, and Andrew Tanenbaum, a professor and the developer of an alternative project known as MINIX. When Torvalds announced his modest Linux project on the MINIX newsgroup, Tanenbaum claimed that the approach Torvalds was pursuing was "obsolete." Torvalds responded by declaring, "Time for some serious flamefesting!" and attacked Tanenbaum and his MINIX and Amoeba projects: "your job is being a professor and researcher: That's one hell of a good excuse for some of the brain-damages of minix. I can only hope (and assume) that Amoeba doesn't suck like minix does."[36] After being told by others that this was not "how it was done," he apologized and signed the message "Linus 'my first, and hopefully last flamefest' Torvalds." But this was far from his last flame.

In response to Kroah-Hartman's posting about the egregious patch, Torvalds wrote "Publicly making fun of people is half the fun of open-source programming. In fact, the real reason to eschew programming in closed environments is that you can't embarrass people in public." In 2013, another Torvalds "flamefest" led some to ask if these outbursts were becoming an embarrassment and liability to the community. The trigger for this incident was Microsoft's requirement that new "Windows certified" computers could run only authorized versions of Windows 8 that were signed by a cryptographic key. Microsoft's stated intention was to make the software more secure, but the change would likely exclude other operating systems from running on those computers. David Howells, a developer from the Linux company Redhat, asked Torvalds to "please" pull a patch that could make it possible to boot Linux on such computers. Microsoft's move was onerous to all Linux developers, but Redhat tried to address it in a way that was not to Torvalds's liking. Torvalds responded that "this is not a dick-sucking contest":

If Red Hat wants to deep-throat Microsoft, that's *your* issue. That has nothing what-so-ever to do with the kernel I maintain. It's your own key. You already wrote the code, for chrissake, it's in that f*cking pull request. Why should *I* care?[37]

This prompted prominent Web developer Evan Prodromou and others to characterize Torvalds's response as sexist and bullying.[38] Flaming and sexist comments are topics that I will return to, but in terms of feedback, this community seems to be harsher than others, perhaps because code is more objective than prose. The performance of a patch can easily be tested, whereas the merit of a fanfic critique is in the eye of the beholder. Also, Linux is a very successful project, and its developers can afford to be picky, as a bad patch can ruin the efforts of thousands of other contributors. Although tweak critique, including patches, is a constructive type of feedback because it is specific and offers a solution, this type of comment is still subject to the vagaries of human ego and emotion.

The Need for "Strong Ears"

Feedback can be distinguished from other types of comment by its intention to help its object (a speaker, student, writer, or coder) achieve a goal. In amateur communities, that goal is recognized as being set by the person who receives the feedback. In its "three Rs of evaluation," Toastmasters stresses that a good evaluation is one in which you "consider the speaker's personal goals." In fanfiction, "There is a difference between expressing dislike for a work and expressing dislike for its premises."[39] "The Art of Critiquing," a guide for *Xena: Warrior Princess* fanfic writers, recommends that beta readers respect that the writer "put a lot of their heart and soul into their work" and "allow the author their story" even if the story fails in the reader's opinion.[40] In software development, commenters will be rebuffed for presuming that they can force a developer to code the features they want. Constructive feedback does not challenge the premise or purpose of someone's efforts.

When it comes to cultures of feedback, I am drawn to the richness of amateur communities. The better Toastmasters clubs could certainly be as effective, if not better, than many public speaking classes taught in accredited universities. Key differences between the amateur and the institutional participant are self-selection and motivation. The amateur is intrinsically motivated, whereas the student or instructor might be there simply for the class credit or paycheck. In a classroom or an office, disappointments must be absorbed as people continue onward—with the help of some griping with peers. In an amateur group, disappointment is easily followed by disappearance, which might be why some amateur communities have well developed norms of concrit. Even so, newcomers and nasties are always arriving, and different groups have varied tolerances for the tone of feedback—from the palatable feedback-sandwich to the bitter medicine of critique or even the poison of complaint.

Whatever its tone, such comment must be digested, and this rarely is an easy task. The Renaissance writer Michel de Montaigne is often described as the father of the essay because he documented his own experiences and thoughts in an unusual (for the time) three-volume collection, *Essais*. He also has been called the world's first blogger because of his self-interest ("I would rather be an authority on myself than on Cicero") and oversharing ("I dislike even thoughts which are unpublishable"). He also strongly identified with his writing: "I am myself the matter of my book." When he presented a copy of his essays to Henri III, the French king was reported to have said that he liked the book, to which Montaigne replied, "Sir, then your majesty must like me." This also implies that criticism of his work was criticism of Montaigne. Hence, the exchange of feedback requires both strength and love: "We need very strong ears to hear ourselves judged frankly; and because there are few who can endure frank criticism without being stung by it, those who venture to criticize us perform a remarkable act of friendship; for to undertake to wound and offend a man for his own good is to have a healthy love for him."[41]

Montaigne's sentiment remains true today. What has changed is the immediacy and accessibility of comment: never before has it been this quick and easy to solicit and give feedback. This is especially so in the case of tweak critique. Furthermore, the distinctions drawn between criticism, feedback, and review sometimes blur, which can be a source of contention within a community. Similarly, the scope and scale of feedback have changed: feedback to a person can be seen by many, and it is easy to encounter unsolicited comment about oneself. Helpful feedback requires care from the giver and emotional work by the receiver. On this latter point, as the rap parody at the start of this chapter requests, we might ask those online to "Be more constructive with your feedback, please."

5

Alienated: You Fail It! Your Skill Is Not Enough!

They call you hater well they're just jealous
Your constructive pearls of wisdom give me thrills I can't deny
How will we know if you don't tell us
We could improve our YouTube channels by "fucking off and dying"? ...
You wished me cancer and misspelled "cancer"
But I know that it's a metaphor. You hope that I will grow,
Just like the tumour you hoped would kill me
Inside the tits on which you said you'd also like a go....
Some might say you're a ...
sexually aggressive, racist, homophobe, misogynistic,
cowardly, illiterate, waste of human skin, ...
But if it wasn't for you my darling,
I would never have written this tune....
—Clever Pie and Isabel Fay, "Thank You Hater"

"Fail." This short comment, frequently seen online, says much: it signifies an ironic, disastrous, or confounding misfortune, with "epic fail" describing the sublimely stupid. This contagious idiom (or *meme*) went viral in 2004 and may have been inspired by a poorly translated caption to a 1998 Japanese video game: "YOU FAIL IT! YOUR SKILL IS NOT ENOUGH."[1] "Fail" appears in comments, in the titles of YouTube videos, and as an "image macro" (a shocking or funny image with a large textual caption). It is the subject of the *FAIL Blog*, which once featured a photo of traffic stopped behind a hearse and its escaped casket. "Fail" is also used self-deprecatingly, as when an online community collectively shakes its head in bewilderment.

"RaceFail '09" is one such incident and resembled a classic flame war from Usenet, the Internet's early discussion forum. On Usenet, a provocative message could spark a heated exchange in which people said things that they ordinarily would not say and later regretted their participation. The metaphor of the flame was apt since tempers flared, and the resulting conflagration spread quickly, often "cross-posted" across newsgroups. The same thing happens on blogs and Twitter today. A friend of mine wished she could opt out of the "retweet fights" that occur when someone in her stream retweets his or her opponent "so that, presumably, we can all fight with them too." In the RaceFail '09 incident, two professional science fiction authors blogged about writing for "the other" (creating characters that are not like the writer). One of those authors, Elizabeth Bear, reflected on the "ongoing problem" of "Writing The Other without being a dick." She recommended that when writers create characters that are unlike themselves, they should think of such characters as people first, listen to others' experiences, research their history, not reduce characters to tokens, avoid stereotypes (especially in creating alien peoples based on our prejudices), and "Accept that no matter what you're doing, some people are going to think you're getting it wrong."[2]

Avalon Willow, a comics enthusiast, blogger, and person of color, thought that Bear had gotten it wrong. In an open letter to Bear, she listed many instances of cultural appropriation and negative stereotypes that she found in science fiction, including some by Bear herself. Although she said that Bear was not a "racist" or a "monster," she thought that Bear's posting demonstrated ignorance and privilege.[3] Friends and colleagues of Bear responded in her defense and a flame war ensued on the issues of covert racism and cultural appropriation. Unlike other cases discussed in this chapter, there was relatively little explicit racism, hate, or harassment. Nonetheless, there were claims of abuse and racism by both sides. Even well-meaning people can get ensnared in angry conversations that leave most participants feeling upset and burnt out. This pattern of discourse exemplifies Godwin's law, an observation that cyberlawyer Mike Godwin made in 1990 about the Usenet: "As an online discussion grows longer, the probability of a comparison involving Nazis or Hitler approaches 1."

However, the understanding of this darker side of online interaction that we inherited from the 1990s, including Godwin's law and the suggestion "Don't feed the trolls," is outdated. Back then, conflict was characterized by participants in an ongoing discussion coming to see their opponents in partisan terms, eventually to the point of calling one another Nazis. This behavior could be sparked or inflamed by a troll seeking to annoy people. Today, comment sections can quickly fill with posts that would've shamed Hitler himself (hence the suggestion "Don't read the comments"). The lone mischievous *troll* who attempted to stir up trouble is now part of a larger culture, and the classic flame wars from before the Web now look harmless with the arrival of *bullies* and *haters*.

Lolz, Trolls, and Anonymity

My middle-school friend Jason used to pester me to hang out on his bulletin board system (BBS), a "war board" that he and his older brother had set up for the sole purpose of trading expletives. As one former BBS user wrote, war boards were "an attempt at witty repartee—a verbal one-upmanship. Unfortunately, because many of us were teens, it was rarely witty and often devolved into 'your mamma' arguments when people ran out of creative ways to insult one another."[4] I never took to it. I could not generate animosity toward people that I had no reason to quarrel with. My BBS of choice was The Science Lab, which was populated by nerds, including some who worked at NASA and the Space Telescope Science Institute. Tempers flared, and insults were exchanged, but that was never the point. In fact, after one such paroxysm, morale plummeted, and some abandoned the system.

Later, when I first experienced the Internet, I recognized elements of both the war board and science board in Usenet's flame wars. Among the hundreds of discussion groups, an exchange could be serious, playful, and heated. This mixture is seen in the famous feud between Linus Torvalds and Andrew Tanenbaum mentioned earlier. Tanenbaum declared that the design of the Linux operating system was "obsolete"; Torvalds responded with "some serious flamefesting" and called Tanenbaum's projects "brain-damaged."[5] Even though some felt that Torvalds had crossed the line, this was different from recent incidents: the hostility was bracketed as flaming, it was limited to the participants, the topic was substantive, ideas were exchanged, and there were no threats or harassment. Some scholars at the time argued that flaming was potentially valuable because it "encourages clear writing and no-nonsense communication": it educated the ignorant, enforced rules, and facilitated effective communication.[6] But an undergraduate is not likely to characterize a professor's work as "brain-damaged" to his face, as Torvalds did to Tanenbaum. This behavior has been confirmed in experiments: online, people exhibit greater status equalization (for example, between student and professor) and disinhibition (such as flaming).[7] One theory is that with a relative paucity of textual interaction, we miss the social cues, context, and information that normally are relied on to regulate interpersonal exchanges. We can easily blunder without realizing how we are affecting other people—despite smiling emoticons.

But what happens if visibility is completely removed? Plato asked this question millennia ago via the story of Gyges, a shepherd who was in the service of the king of Lydia. Gyges found a ring of invisibility and used it to bed the queen, kill the king, and become the new ruler. Plato asked whether a just man could be corrupted in such a circumstance. J.R.R. Tolkien thought so and had Frodo, a virtuous and modest hobbit, falter at journey's end and fail to cast the One Ring into the fires of Mount Doom. (It is destroyed by Sméagol, who bites off Frodo's finger to reclaim his "precious" and accidentally falls into the fires below.) More recently, in an episode of the radio program *This American Life*, people were asked if they would prefer to be an invisible man or a hawk man. The consensus seemed to be that flight would likely lead people along a heroic path and invisibility would lead to shoplifting and voyeurism.

These philosophical and popular suppositions have empirical support. In 1969, psychologist Phil Zimbardo reported an experiment in which people were asked to administer shocks to others. Research accomplices then pretended to receive the shocks. Researchers found that participants who wore large lab coats and hoods were more willing to shock others than participants who wore name tags. Zimbardo believed that the veiled subjects experienced *deindividuation*: a loss of a sense of self and social norms. In another early study, thirteen hundred children were secretly observed trick-or-treating. They were told to take a single candy. Lone children who were identified by name rarely cheated, and anonymous children in a group cheated most of the time.[8]

More recently, psychology researcher Tatsuya Nogami attempted to tease apart the differences between identity (such as a name) and anonymity (an inability to associate a person with his or her behavior). As part of a take-home assignment, he asked over a hundred university students to flip a coin twice. Some students were asked to identify themselves on their coin-flipping report, and some received a reward (a coupon book) for getting two tails, which should happen 25 percent of the time. Nogami considered all subjects to be anonymous since he could not know if any individual cheated. Instead, he could infer cheating based on the aggregate statistics. Of the nonidentified subjects who could get a reward, 46 percent reported flipping two tails—well in excess of the expected 25 percent and evidence of cheating. He was surprised to find that those who had identified themselves (but still could not be associated with cheating) did not choose to cheat: only 21 percent reported getting two tails. Those in the latter group were just as "anonymous" and unaccountable as those of the first group because no individual could be blamed: any single person might easily get two tails. Nogami concluded by suggesting that asking people to identify themselves perhaps prompted them to be more aware of their ethical standards even when they could not be linked to unethical behavior.[9] Although they often overlap, a sense of self and a sense of accountability can be distinct.

But both can easily be obscured online. In the 1990s, scholars suggested that because the media experience was not as rich as face-to-face interaction, social cues could be lost, and people could have a deindividuated sense of themselves. Popular theories also arose. The Urban Dictionary defines *Internet balls* as the courage to use a computer screen to write "whatever you want, to whomever you want" in a way you would not if you were face to face: "That would require having balls when you're away from your computer ... which you don't." Similarly, the "Internet fuckwad theory" posits that "a normal person + anonymity + an audience = total fuckwad."[10] Some have worried that even people who participate online as themselves might become more extreme by placing themselves in an online bubble of the likeminded or by being exposed to mean-spirited comments that make them more polarized in their views.[11]

The potential for misunderstanding and rancor online (as seen in Race-Fail '09) is a tempting target for pranksters. *Griefing* emerged in online games where some players would annoy and harass others rather than play to achieve the putative goal of the game. Some killed members of their own team. On Usenet, trolls provoked newcomers with outlandish statements that were likely to incite flame wars. The term *troll* was likely

THE HISTORY OF SOCIAL

CHAPTER 1:
IN 1985 BEING A TROLL WAS
NOT THAT MUCH FUN

borrowed from the notion of trolling for fish with baited lines; the aphorism "don't feed the trolls" advises that one should ignore "flame bait."
As the "Troller's FAQ" notes, "If you don't fall for the joke, you get to
be in on it."[12] Linguist Susan Herring was one of the first to study such
behavior in the 1990s. In one early article, she detailed the tactics and
consequences of a particular troll in a feminist forum. From her reading
of trolling culture, Herring defined a troll as one who sends messages
that appear outwardly sincere, that are designed to attract predictable
responses or flames, and that consequently waste time by provoking futile
arguments. In the case she discussed, "Kent" refused to acknowledge others' points, willfully misinterpreted others' motives and views, attacked
others for ignoring him, and implied that he could change if people would
simply explain their concerns to him. He continued to bait those who
did engage him and taunt those who would not. The community was

in a bind: if it permitted him to stay, its forum would be polluted with his harassment, but if it banned him, it risked being labeled censorious. Moreover, beyond the personal distress that trolls may cause, communities can become brittle as members argue about what to do or become more paranoid, especially toward newcomers.[13]

In hindsight, the trolling of the 1990s can seem relatively innocuous. Anthropologist Biella Coleman has noted that "Trolls have transformed what were more occasional and sporadic acts, often focused on virtual arguments called flaming or flame wars, into a full-blown set of cultural norms and set of linguistic practices."[14] This shift is also seen in a competing theory to deindividuation. Instead of (or in addition to) a loss of self and abandonment of social norms, perhaps social media lead people to experience *depersonalization*, a shift from a sense of self toward a group and its norms.[15] Evidence for this theory can be seen as far back as 1979, when a pair of researchers wondered if Zimbardo's finding about hooded experiment participants had more to do with the hoods than with the anonymity. They replicated the experiment by dressing two groups of women in outfits like those of the Ku Klux Klan and nurses. Those who wore the KKK-like outfits were more aggressive in administering shocks than those who dressed as nurses.[16] That is, the subjects did not abandon all norms but adapted to the norms associated with their dress. Zimbardo himself demonstrated the power of context and uniforms in a prison experiment in which those who dressed as police soon became autocratic.[17] Today, some take their cues from a culture of trolls and griefers in search of laughs (or *lolz,* taken from "laugh out loud") or in the expression of hate.

All of these theories might be characterized as "good people acting badly" explanations, which focus on the lack of feedback from the recipient of communications. Absent a rich media, social cues are filtered out, social presence is attenuated, and people do not appreciate their effects on others. If people could see that they upset someone, then most would be less likely to do so. Other "good people" theories relate to the originator of the communication. Under deindividuation, we lose sense of ourselves and inhibitions. Under depersonalization, morality shifts toward a different set of norms.

However, there are also theories and cases of what I call "bad people acting up." In this view, some of what is seen online is the disproportionate effect of a difficult minority. In a 2014 study, over a thousand respondents completed a personality inventory and survey of their online commenting. Only 5.6 percent of those surveyed reported that they enjoyed trolling, yet there were strong positive relationships between the expressed enjoyment of trolling, measures of sadism, and a high frequency of online commenting.[18] At the extreme, there are frightening folks like Luka Magnotta, who was diagnosed with paranoid schizophrenia in his teens. In his twenties, he became notorious for his online exploits, including suffocating cats and killing and dismembering a man—all posted on YouTube. In 2012, he fled to Europe, where he continued posting videos taunting police and thanking "his fans" for their attention and support. Magnotta was eventually arrested in an Internet café in Germany reading stories (and likely commenting) about himself. Few people have such a disorder or encounter anyone who does, but online everyone is just a click away, and it can be difficult to know who is writing the comments.

Despite the shift in our conception of trolls as a lone man like "Kent," trolling still appears to be the province of men. In the 2014 trolling study discussed above, men were more strongly associated with trolling than women.[19] Whitney Phillips, writing about those "LOLing at tragedy" (those who troll Facebook memorials), has noted that "Facebook trolling, like trolling generally, is an absolute sausagefest." In two years and dozens of interactions, she encountered "a mere handful" of trolls who identified as women. And those who identify as female are not necessarily so: they "tend to affect and in some cases even amplify the same sexist posture and language as their male counterparts" or use their (supposed) femininity to "accomplish some unholy objective."[20] Furthermore, this gendered pattern persists when it comes to actual harassment. In her book *Hate Crimes in Cyberspace*, law professor Danielle Keats Citron noted that although men are targets of cyberharassment, it is "beyond dispute" that "being a woman raises one's risk of cyber harassment" as may being a women of color or lesbian, bisexual, or transgender: "Nearly all of the victims I've talked to have been female."[21]

This discussion of trolls might lead one to think that the line between those acting in good faith and mischievous trolls is a clean one: that if one simply acted sensibly and followed the aphorism "don't feed the trolls" one would be safe. But it is no longer so easy.

Mean Kids, Mean Men, and "FREEZE PEACH"

According to its founders, the now-defunct group blog *Mean Kids* was intended to be a forum for "art and criticism, pointed and insulting satire"; it considered itself to be a place of "purposeful anarchy" with a tradition of "you own your own words."[22] This forum for unmoderated insults soon led to misogynistic threats. In March 2007, Kathy Sierra, author of several Java programming books and a popular blog, wrote that she had canceled her workshop and keynote speech at a conference and instead was at home "with the doors locked, terrified":

> It began just over four weeks ago, when something shifted. It started with death threat blog comments left here (which some of you may have seen before I deleted them) including: "Comment: fuck off you boring slut.... i hope someone slits your throat and cums down your gob." We all have trolls—but until four weeks ago, none of mine had threatened death.... At first, it was the usual stuff—lots of slamming of people.... Nothing new. No big deal. Nothing they hadn't done on their own blogs many times before. But when it was my turn, somebody crossed a line. They posted a photo of a noose next to my head, and one of their members (posting as "Joey") commented "the only thing Kathy has to offer me is that noose in her neck size."[23]

As the story received greater attention, the attacks escalated. In addition to being a (distressing) milestone of sorts, exposing this facet of online culture to a wide audience through CNN, the BBC, and the *New York Times,* it showed how trolling had metastasized. Trolls had always sought to provoke a response, but writing offensive and hateful comments had emerged as a creature of its own. *Haters* try to upset and belittle others by expressing extreme hostility and attacking any aspect of a person that is likely to cause distress (such as gender, ethnicity, sexuality, and appearance). The widespread use of the term *hater* likely began with the expression "Haters gonna hate" from hip-hop culture. Much like the warning "Don't feed the trolls," it implies that some kinds of negativity are best ignored.

But ignoring no longer seems sufficient when faced with a hater. One reason for this is that the labels of *troll, hater,* and *bully* have lost some of their descriptive potency. Today these terms are loosely bandied about in arguments, and some people use them for anyone who disagrees with them. More substantively, the hate expressed online today has a frighteningly sharp edge and long reach. For instance, hateful speech is magnified via disturbing images and videos. The manipulation and use of images, such as macros and animated GIFs, is common in Internet lolz culture. Although these GIFs are often funny, in the hater context they can be alarming. Sierra, for example, was pictured as muzzled with women's underwear. Other targets of haters have received gruesome images. Also, privacy is often violated through the practice of "doxing," or publicly *documenting* the target's contact, financial, and health information. Haters cannot be ignored when they make threatening phone calls, including to family members, friends, and employers. The *Mean Kids* incident was more than a flame war in which teenagers bandied insults about one another's mothers. Also, the fact that the threats began with the hostility of notable people made it all the more distressing. This is a feature of what I call a *trollplex*: an attack by people who come from varied backgrounds, exhibit varied behavior, but share a target, a culture, and venues.

In this case, the trollplex included well-known bloggers Chris Locke (who blogged under the name "rageboy") and Frank Pynter. Locke and Pynter ran *Mean Kids* and contributed to its overall tone (Locke called Sierra "a hopeless dipshit"),[24] but they made no threats. It also included *Mean Kids* contributor "Rev Ed," who posted Photoshopped images of Sierra. (When the person behind the "Rev Ed" account was exposed, he claimed that his computer had been hacked and he had not posted the materials.) Others, whose identities were never publicly revealed, made explicit threats. The attacks on Sierra included insults, frightening threats, and harassment, but all melded into a single discourse that was rooted in the discussion at *Mean Kids* and other blogs. Some criticized Sierra for condemning all who contributed to the site, but she maintained that participants had a responsibility for creating an environment for this type of culture and speech.

The increasing attention that resulted from the story was unwelcome to many. Things were getting worse for Sierra as others joined in on the harassment, the identifiable *Mean Kids* contributors were embarrassed, and those who were not yet identified feared exposure. *Mean Kids* and the related blog *Unclebobism* were removed, and in a surreal attempt to end the incident, Sierra and Locke issued a joint statement and appeared in a televised meeting on CNN. In the joint statement, Sierra wrote that although she did not feel that the *Mean Kids* proprietors were responsible for the threats, they still had their differences:

However, Chris and I (and others) still strongly disagree about whether people who are respected and trusted in our industry ... are giving tacit approval when they support (though ownership, authoring, and promoting) sites like meankids and unclebob. This is about trust and leadership in our community, and whether those who are looked up to have a (non-legal) responsibility to the community whose trust they've earned for the things they promote.[25]

Beyond the history of animosity that people at the *Mean Kids* blog felt toward Sierra and others, she suspected that the trigger was her support for bloggers who delete inappropriate comments from their own blogs. This might seem commonsensical today (although it still prompts anger in some), but it was a more controversial position then. Also, the conflagration was likely related to her admission of fear and her style of writing, which differed from a stereotypically masculine norm. The reason that the advice to "Ignore the trolls" has currency is because a panicked or fearful response to negative comments encourages more trolling. Fishing metaphors abound. Reactions to trolls and threats are said to be like "chum in the water." One of the sharks that was attracted by Sierra's distress was the infamous Internet troll "weev," who, among other things, revealed Sierra's social security number and home address and invited others to "send them gifts that properly express your sentiments." Sierra later noted that "People did. We moved." weev explained the harassment by writing, "Kathy hollers like a stuck pig as she wonders why the trolls escalated to magnitudes which she could no longer control. The answer is obvious: she fought the LOL. The LOL won."[26] Although those who left snarky comments saw themselves as distinct from those who left threats (which can be direct or indirect), such nuances are understandably lost on the frightened person at the center of a trollplex.

There is also the issue of gender. Susan Herring's study of trolling was preceded by work on the gender dimensions of flaming. In 1993, she reported that on the lists that she studied, only about 5 percent of the posters, nearly all of whom were men, were responsible for most of the adversarial rhetoric. (They also tended to dominate in the number of words written.) This led her in the following year to ask, why do "women thank and men flame?" Given that flaming is usually the behavior of a minority of (mostly) men, she discounted simple disinhibition.

Her original hypothesis was that perhaps men and women felt differently about politeness; however, both groups reported valuing politeness and disliking rudeness. She concluded that men had an overlapping but dominant value system: men assigned "greater importance to freedom of expression and firmness of verbal action than to possible consequences to the addressee's face needs." These men flamed to "regulate the social order" in accordance with these values "as self-appointed vigilantes on the 'virtual frontier.'"[27]

Locke's handle of "rageboy" certainly evokes the persona of an angry (juvenile) male. Moreover, in the joint statement, while he condemned the "offensive words and images," his main concern seemingly was about free expression. He concluded his statement with a warning that the U.S. Constitution's first amendment protects speech "that many find noxious" and we must be wary of forces in the world "that would leap at any opportunity to limit speech or even abolish certain forms of it. Crucial as is the current debate about hate speech directed at women, it would be tragic if this incident were used as a weapon by those who would limit free and open exchange."[28] However, the first amendment prohibits the U.S. Congress from abridging speech (which was not suggested in this case) and says nothing about what individuals, organizations, and communities can condone or condemn. Nonetheless, noxious speech is often justified by way of anarchic and libertarian rhetoric about freedom. As Andrea Weckerle, author of *Civility in the Digital Age*, has noted, this focus on freedom often makes "innocent victims appear as though they are opposed to freedom of speech, when actually they are opposed to lies and injury."[29] This is reflected in Sierra's statement in which she was obliged to qualify that her "desire is for much more open debate on this issue, not legislated limits." Amusingly, "freedom of speech" is so frequently invoked in defense of offensive and harassing speech that it is now popularly parodied by the exclamation "FREEZE PEACH!" This phrase is used to describe "whiny, entitled behavior" from those "misogynists who think that FREEZE PEACH! means their right to pester women in any way they choose or use any kind of misogynist language they see fit is sacrosanct."[30]

One *Mean Kid* contributor maintained that Sierra's style invited abuse, to which she then overreacted, exemplifying Herring's findings about the gendered regulation of speech. He wrote that Sierra "mixes a type of new-age rhetorical spirituality with computer science." He believed that her aphorisms of "code like a girl" and "beauty drives the computer industry" deserved ridicule. Sierra's pretense of technical chops was belied by her alleged ignorance of Internet meme culture, as when she overreacted to the comment "IMMA KILL YOU," which was "borne of Japanese anime and for those who know, it is also hilarious." (This particular meme was featured in a 2010 Judd Apatow film in which the character played by Jonah Hill receives a threatening text message: "Where the fuck are you? Imma kill you. Smiley Face.") The whole incident "was a strange collection of odd synergies mixed up with childishness and, frankly, fun between people I was enjoying interacting with." Finally, "I must add this: authors who write with less childlike magical thinking might also find they receive less childish criticism of their works."[31]

Such justifications for abusive behavior often show what psychologist Albert Bandura has identified as "moral disengagement." People try to lessen the cognitive dissonance of seeing themselves as decent people who do indecent things by using justification ("she deserved it"), euphemistic language (just "trolling" or "having fun"), and advantageous comparisons ("I never threatened her"). They dehumanize the target ("a stuck pig") and disregard or misrepresent the injurious consequences of their actions ("she needs to toughen up"). And they displace or diffuse responsibility by saying they were only a small part of the conduct. Such is the morality of the trollplex. Yet as Bandura noted, "People suffer from the wrongs done to them, regardless of how perpetrators might justify their inhumane actions."[32]

Dongles, Anonymous, and Kneejerks

In March 2013, a "mr-hank" posted a comment on the popular geek discussion site Hacker News in which he concluded: "Let this serve as a message to everyone, our actions and words, big or small, can have a serious impact."[33] This observation about comment was part of an apology and explanation of his involvement in what became known as the "forking and dongle incident" at PyCon 2013, the annual conference dedicated to the open-source Python programming language. The conference had well over a hundred sponsors, sold out its 2,500 registrations, and 20 percent of the attendees were women. These figures reflect Python's reputation as a popular language with a friendly community. In light of a history of inappropriate presentations and sexual harassment at similar conferences, PyCon also had a code of conduct: "Sexual language and imagery is not appropriate for any conference venue, including talks.... Behave professionally. Remember that harassment and sexist, racist, or exclusionary jokes are not appropriate for PyCon."[34] This code was at the center of a widespread debate, sometimes referred to as "Donglegate," about the appropriateness, context, and geography of comment in the age of the Web. (A dongle is a small electronic device, like a USB key, that is inserted into and extends the capabilities of a larger device.)

On her blog, Adria Richards described herself as an "excessively enthusiastic technology evangelist," and she was at PyCon 2013 on behalf of her employer and conference sponsor SendGrid. On March 18, the day after the main conference ended, Adria Richards blogged, "Yesterday, I publicly called out a group of guys at the PyCon conference who were not being respectful to the community." She explained that during the closing plenary session, mr-hank (known only by a pseudonym) and a colleague were sitting behind her and kibitzing about the day's sessions. The two were employees of PlayHaven, which also was a conference sponsor. mr-hank noted that he did not find much value in an earlier session, and Richards turned around to agree with him: "He then went onto say that an earlier session he'd been to where the speaker was talking about images and visualization with Python was really good, even if it seemed to him the speaker wasn't really an expert on images. He said he would be interested in forking the repo and continuing development."

"Forking a repository" is a common open-source practice in which a project is copied and extended (creating a *fork* in the history of the code) and can be merged back into the original or continued independently. Richards noted, "That would have been fine until the guy next to him began making sexual forking jokes." Richards had already dealt with a much more inappropriate sexual joke earlier in the day when a developer spoke to her about his failed attempt at humor with another woman. Unlike then, she now was in a "packed ballroom" and could not easily discuss the matter. Annoyed, she turned her attention back to the conference organizer:

Jesse Noller was up on stage thanking the sponsors. The guys behind me (one off to the right) said, "You can thank me, you can thank me."... They started talking about "big" dongles. I could feel my face getting flustered. Was this really happening? How many times do I have to deal with this? Can they not hear what Jesse is saying? ... I was telling myself if they made one more sexual joke, I'd say something. Then it happened: the trigger. Jesse was on the main stage with thousands of people sitting in the audience. He was talking about helping the next generation learn to program and how happy PyCon was with the Young Coders workshop (which I volunteered at). He was mentioning that the PyLadies auction had raised $10,000 in a single night and the funds would be used for their initiatives. I saw a photo on main stage of a little girl who had been in the Young Coders workshop. I realized I had to do something or she would never have the chance to learn and love programming because the ass clowns behind me would make it impossible for her to do so.[35]

Richards considered her options and, during a change of speakers, stood and took a photo of the two men behind her. She then tweeted the photo to those following the #pycon hashtag: "Not cool. Jokes about forking repo's in a sexual way and 'big' dongles. Right behind me." Her nearly two thousand Twitter followers could also see the tweet. A few minutes later, she tweeted "someone talk to these guys about their conduct." Conference organizers separately spoke with Richards and then the two men privately. The men agreed that the comments were in poor taste and left of their own accord. The conference was ending anyway. The staff tweeted that the incident had been addressed, but it was not yet over. The story gathered increasing attention online, especially at the popular discussion sites Reddit and Hacker News. Richards posted a statement on her blog explaining her perspective, as did the conference organizers. mr-hank started a new thread at Hacker News in which he apologized for the dongle joke, disclaimed any sexual innuendo in the discussion of "forking," and noted that he had been fired:

Hi, I'm the guy who made a comment about big dongles. First of all I'd like to say I'm sorry. I really did not mean to offend anyone and I really do regret the comment and how it made Adria feel. She had every right to report me to staff, and I defend her position.... My second comment is this, Adria has an audience and is a successful person of the media. Just check out her web page linked in her twitter account, her hard work and social activism speaks for itself. With that great power and reach comes responsibility. As a result of the picture she took I was let go from my job today. Which sucks because I have 3 kids and I really liked that job. She gave me no warning, she smiled while she snapped the pic and sealed my fate. Let this serve as a message to everyone, our actions and words, big or small, can have a serious impact.[36]

Others were not so charitable, claiming that mr-hank had nothing to be sorry for: that this was political correctness run amok, goofy sexual puns are not necessarily sexist, and Richard's public shaming of the two men was wrong. Other details of the story emerged. Only mr-hank was let go and not his colleague, which implied that other factors might have been involved in his dismissal. (On Reddit, Richards thanked mr-hank for his message and hoped that his former employers would reconsider their decision.) People also began to scrutinize Richards and accuse her of hypocrisy and attention-seeking. Although some might have felt that Richards was overly sensitive, there have been dozens of reports of sexualized presentations and sexual harassment at recent technical conferences. Any particular incident that gains widespread attention may be only the most recent (and relatively innocuous) one in a lifetime's worth of subtle bias and overt abuse. As noted, Richards had dealt with a more egregious incident earlier in the day. Additionally, as blogger Gayle Laakmann McDowell wrote, Richards had responded in all of these cases "in a fairly reasonable way. She was not overly aggressive or hostile. Rather, she explained her objections clearly and fairly."[37] The same cannot be said of the response toward Richards.

As the dongle incident gained attention online, attacks were made on Richards, the conference organizer Jesse Noller, and Richards's employer. Richards, a woman of color, was called a cunt, bitch, and nigger and threatened with rape and murder. As has been noted, sexually violent comments, especially toward women, are an established genre of comment. Media scholar Emma Jane has noted that such "e-bile" is characterized by profanity, ad hominem invective, stereotype, and hyperbolic imagery of graphic (and often sexualized) violence that manifests as a threat or wishful thinking.[38] For instance, tweets to Richards told her to "kill herself," "shut the fuck up stupid bitch and go to the kitchen," and "you need to be gang raped so you get some common sense." Her contact information was published alongside threats, including a gruesome image of a beheaded, gagged woman accompanied by the text "when I'm done."

The worst of these comments likely arose from two sources. First, there is an online community of "men's rights" activists, some of whom vent wildly misogynistic speech. Second, trolls have long been provocateurs but they now can be much more vicious. A classic but relatively innocuous Usenet trolling from the 1990s was entitled "Oh how I envy American students." In the post, the author spoke of his admiration for fraternity life, its drinking, and its "para-homosexuality"; this bait prompted over three thousand comments. This pales in comparison to the 2008 infiltration of the Epilepsy Foundation's message boards with images that were intended to trigger seizures.[39] Most trolls today are likely too young to remember Usenet and instead trace their roots to 4chan. This "imageboard" originally focused on Japanese pop culture and today is famous for the "random" board where funny, weird, and gross images are posted and commented on anonymously. On a single day, tens of thousands of threads (posts with their corresponding comments) flow through the board. On average, they last only seconds on the front page and for minutes on the subsequent pages. As anthropologist Biella Coleman noted, "These rapid-fire conditions magnify the need for audacious, unusual, gross, or funny content."[40]

Despite the frivolity of many 4chan discussions and the insincerity of many trolls, 4chan was also an early home to the Anonymous "hacktivist" movement that is associated with online pranks and protests, especially against the Church of Scientology. A group identifying itself as Anonymous targeted Richards and her employer. Both her site and Sendgrid's became unreachable during a denial-of-service (DoS) attack. Such an attack bombards a site with bogus requests, often from infected computers, making it difficult for it to serve genuine users. A petition demanded that she be fired, and a manifesto was anonymously published, noting that "the outrage over the petty and malicious conduct of your employee, Adria Richards, is about to erupt in an explosion of lulz and collateral damage over anyone and anything that had the misfortune of being in contact with this individual." Harassment against Sendgrid's customers, investors, and employees was threatened, including, as the manifesto enumerated, "obnoxious phone calls, emails, denial of service attacks, online vandalism and defamation, and even real-life harassment."[41] Anonymous arose, in part, out of the culture and anonymity of 4chan; they use these tactics against targets whom they feel engage in hypocrisy, secrecy, censorship, and corruption. Sometimes their activities seem incongruent: in the same period that Anonymous was threatening harassment of Richards, others, also claiming to be Anonymous, were exposing a cover-up of high school football players who sexually assaulted a teenage girl. (The boys had posted images of and comments about the assault online and Anonymous excels at ferreting out such material.) To the extent that the morality and motivation of Anonymous's members can be understood, their ethic is often a reflection of their tactics rather than the other way around. That is, almost anyone can claim the Anonymous mantle, and their activities are shaped more by the tools of exposure and harassment than by a consistent moral philosophy.

In the "dongle incident," the threats of exposure and harassment were followed later in the day by a blog post from Sendgrid's CEO stating that "Effective immediately, SendGrid has terminated the employment of Adria Richards. While we generally are sensitive and confidential with respect to employee matters, the situation has taken on a public nature. We have taken action that we believe is in the overall best interests of SendGrid, its employees, and our customers." He later explained that although the company supported the right to report inappropriate behavior, Richards's public photo and tweet crossed the line: she "can no longer be effective in her role" as an evangelist. He wrote that she had "divided the same community she was supposed to unite" and her approach put the business, its employees, and its customers in danger.[42]

Many people who read about this story probably thought to themselves: "I've made similar or worse stupid jokes. Do I deserve to be fired?" Yet the intangibility of comment prompts anxieties and conflicts related to place. Is a geek conference a social venue or a professional space? In a conference session, is a comment that can be overheard still just between friends? Does tweeting a photo of conference participants violate a boundary between private and public? This story also received a lot of attention because of the character of comment itself. Comment's shortness, asynchronicity, and reactivity mean that it is often highly contextual, but the link between a comment and its context is also easily lost. Because it is *hypotextual* (as described in the previous chapter), misunderstandings easily arise, and everyone else can comment on these conflicts without reading much background or considering context and nuance. As soon as a story is posted on Reddit, thousands of kneejerk comments might follow. We all have opinions, and many do not hesitate to express them at the bottom of the Web—even when they are less than informed ("tl;dr").

Goodreads and Bully-Battles

As discussed in an earlier chapter, the relationship between authors and reviewers can be testy. It was hoped that at Goodreads things might be different. Goodreads is a social networking site where users review, rate, discuss, and make lists of books and even interact with their favorite authors. Goodreads encourages authors to promote themselves and their books by developing a profile or blog, sharing excerpts of new works, and participating in discussions about their books. Many users self-identify as both avid readers and hopeful authors. However, the site gained a different sort of reputation, especially with the launch of the Stop the Goodreads Bullies (STGRB) Website in July 2012.

Professional authors and critics might complain about one another, but their established careers and public identities tend to keep their exchanges relatively sane. For amateurs who write fanfiction as a hobby, snark can be hurtful and provoke a reaction, but their living is not at stake, and readers are not able to complain of wasted money. Conflict is most heated among independent publishers and their readers. Self-published authors who wish to make a living from writing intersect with readers who have paid for works of uneven quality. STGRB set out to identify commenters on Goodreads who worked "to destroy that author's reputation and career for either their own personal amusement or for vengeance."[43] However, in the exchange of accusations and recrimination, it was difficult to discern who was bullying whom.

To begin to untangle this tale, consider how the Web has changed publishing and reading. Foz Meadows is a member of the young adult (YA) fantasy community and is a self-described geek who is fond of silly hats. She reads, reviews, and writes urban fantasy fiction and blogs at the *Huffington Post*. When she was younger, she felt that reviews were never about books she might like, but this changed with the Web. Online forums and real life science fiction and fantasy (SFF) conventions exposed her to a wider community and selection of authors. Amazon and eBay made it possible to buy books that she could not find in a local store. Digital publishers like Lulu helped independent authors produce and distribute works that would never have found an audience. And book bloggers and sites like Goodreads made it possible to find likeminded readers and their "good reads" recommendations. She wrote, "I regularly buy not only ebooks, but firsthand and secondhand books online. I keep a list of titles that have caught my interest on Goodreads, and a wishlist of books I plan to buy on Amazon."[44] In this marketplace, where many readers are probably young (teens to thirties) and digitally capable, authors are well served and might even enjoy discussions with their readers. These are all positives for online comment.

This is also a marketplace in which an author's success is affected by online reviews. Such reviews can be abused by both friends and enemies who mete out praise or punishment. (On the photography-sharing site photo.net, such partisan behavior is referred to as "mate" and "revenge" rating.) Those behind the STGRB Website felt that some readers on Goodreads bullied authors by insulting them, posting malicious reviews, and placing books not yet released on "do not read" lists. One STGRB-affiliated author posted his own list of bullies: "So here are the bullies for all of you to see. People who have attacked me and continue to do so by rating and reviewing my books without even reading them." The titles of peoples' do-not-read lists seemed especially irksome: "not-in-a-gazillion-years," "I-refuse-to-read," and "crazy-shit-authors-to-avoid." This prompted him to conclude, "They just hate me because they are haters. And bullies."[45] I consider such lists to be a *drama genre* of comment (much like the *question and answer genre* that is discussed in the next chapter.)

Social media scholars Alice Marwick and danah boyd have noted that "Drama is the language that teens—most notably girls—use to describe a host of activities and practices ranging from gossip, flirting, arguing, and joking to more serious issues of jealousy, ostracization, and name-calling." Key components include that it is relational and social, reciprocal, and often performed for and magnified by an online audience.[46] *Drama genres* of comment lend themselves to this type of practice. Interestingly, while drama can overlap with bullying, teens tend to avoid the latter term because it is tainted with adult consternation and a presumption of victimhood. Adults seem to be more likely than teens to claim being the victim of online bullies.

Another reason for the drama at Goodreads is that the snark being exchanged between readers at Goodreads is visible, almost unavoidably so, to the authors. As noted earlier, in the age of the Web, unsolicited and unwelcome comments can easily find their way to anyone. Sporkers, the snarky commentators of amateur fanfiction, learned this lesson: if you wish to avoid angry authors, don't link your spork back to them; what authors don't know won't hurt them. Unfortunately, Goodreads was not so savvy. According to Meadows's analysis of what went wrong, author and reader spaces overlapped. Part of the controversy was about which audience Goodreads really served. It advertised itself as a social site for "readers and book recommendations" but at the same time it promoted itself to authors with a different message: "With an audience of more than 10 million affluent, educated, self-identified readers, Goodreads is the ideal place to advertise your book."[47] The initial, perhaps naïve, notion was that both readers and authors could happily inhabit the space. Perhaps they could if there were no mean reviewers or defensive authors, but such an assumption is equally naïve. As Meadows noted, the architecture of the site made subsequent conflagrations all but inevitable: "it's analogous to carrying on a bitchy conversation within earshot of the person you're talking about, with the added rider that, as this is also a professional space for the author, they're not allowed to retaliate—or at least, they can do so, but regardless of the provocation, they'll come off looking the worse. Which leads to fans—and, sometimes, friends—of authors leaping to their defense, often with disastrous results."[48]

In an essay on the *Huffington Post*, an anonymous STGRB author noted that a bullied author's book might receive as many negative reviews in a single day as it had previously accumulated in total: "The next day, the same group of bullies all review another book that is currently on their target list. This is not a conspiracy. They are very open about which book they are currently targeting, and the bully reviews always coincide with their current hate-list."[49] STGRB aimed to document the abuse: "Our methods are extreme but after trying to reason with the bullies, after trying to appeal to their humanity, all of which failed, we have come to understand that only by outing them, can we get their attention." That is, bullies must be held accountable, and the toxic atmosphere of Goodreads remedied. STGRB advocates argued that much of this would be achieved if Goodreads policed abuse in its forums and enforced the rule that reviews must be about books rather than authors.

On first reading, this seems like a reasonable position, but many Goodreads participants objected to the tactic of "outing" (and perhaps endangering) readers via an anonymous site. The badges and banners of antibullying organizations were removed from STGRB when those organizations learned about the site's tactics. The resulting controversy about the essay prompted Andrew Losowsky, the *Huffington Post*'s books editor, to explain why the essay had been posted in the first place. He did not apologize or offer to take the piece down because it abided by the *Post*'s code of conduct and he felt that the authors were entitled to their say, but he did follow it with a piece by Foz Meadows. She and others noted that there was evidence of poor behavior on all sides, but she argued that disagreement, unkind reviews, and public insults are not synonymous with bullying. She claimed that the notion is being abused, such as when politicians claim that they are bullied when others challenge their sexist or racist remarks. For her, bullying requires an imbalance of power through which one person harasses, belittles, and undermines a weaker person. According to Meadows, real bullying attacks are "vicious, personal, and often constitute criminal offenses … a situation which is demonstrably not the same as some snarky, unpaid reviewers slagging off a book." She did not consider the sexist comments left on her blog to be bullying, either. Although they are offensive and abusive, she has the power to delete comments, ban commenters, "and publicly mock them for their opinions."[50]

However, I do not find the boundaries of online bullying to be as clear cut as Meadows. This case is veiled in a fog of confusion that often results from what I refer to as a *bully-battle*. The authors of the 2012 book *Cyberbullying* wrote that while bullying can be understood as intentional aggressive behavior that involves an imbalance of power, cyberbullying is a bit more complex.[51] Does an anonymous commentator have any power? Does the definition apply to adults as well as to minors? Can both sides be both victims and perpetrators? (I would answer yes to all of these questions.) Indeed, the confusion resulting from this later point, of reciprocal accusations ("you are a bully," "no, you are!"), is the first characteristic of a bully-battle. Amidst the background noise of clearly abusive comment (much, but not all, of it from trolls) the partisans argue about the implications of veiled warnings, the ethics of collecting public information (such as that found on Facebook), exposing identifiable information (like addresses and phone numbers), and the making of amateur diagnoses (including that someone is "obviously a psychopath"). Reciprocal accusations can even include former friends, colleagues, and romantic partners.

Beyond reciprocal accusations, a second characteristic of the bully-battle is division of the community and partisanship. People on one side of an issue characterize those on the other side as bullies and seek to correct perceived injustices by aligning with their allies in a conflict of "us against them." Third, the conflict is characterized by a strident rights-based rhetoric. For instance, bullied authors complain of being "censored," and bullied readers declaim that "readers have rights," as one anti-STGRB site is named.[52] Fourth, identities and personal information are revealed or "outed" in a practice known as "doxing" or "doc dropping." People opt to use pseudonyms online for many reasons: people want to keep their hobbies private from the scrutiny of their employer or avoid online stalkers. They might also use anonymity to behave in a way that others object to, hence doxing is intended to expose hypocrisy and make accountable those who were anonymous; it is also used to harass. Doxing is clearly problematic, as Meadows wrote about the Stop the Goodreads Bullies site: "Without this single aspect, the site itself—despite its occasionally libelous descriptions—might not have attracted nearly so much attention. With it, however, the entire purpose moves from being a counterculture to something legitimately sinister, with various reviewers not only finding their photos, real names and cities of residence posted, but also the names of their partners, employers and—in at least one instance—the name of their family's favorite restaurant together with the days and times they usually frequent it."[53] Doxing is often practiced by way of "watch" forums, blogs, and hashtags in which suspect behavior is exhaustively collected and discussed.

Fifth, and finally, lists of opponents, often doxed, are drawn up and publicly posted. A useful feature of Goodreads is that members can create lists or "shelves" of works. This feature initially was used to collect works that a member recommended or hoped to read, but it became a way to censure others. For instance, one reader wrote of an author who "became offended when she found her book shelved as something I did not wish to read. She felt bullied … because I did not want to read her book after she applauded my stalker."[54] This is then reciprocated: after "won't read" lists were created by some readers, STGRB authors drew up lists of "bullying" readers; these authors (in turn) were subsequently exposed on watch sites, including the blog *Badly Behaving Authors*. Goodreads, alarmed by the acrimony and bad publicity, has attempted to curb some of the rancor by asking its readers to focus on books rather than authors and declaring some comments "off topic." This turned some partisans against the Goodreads site itself, with combatants reposting deleted reviews and promiscuously labeling others' comments as off topic.

This pattern of behavior is not an isolated incident, and the bully-battle pattern can be found in various communities, including one that prides itself on its rationality.

Bright Bullies and Bashtags

The "free thought" community thinks itself judicious, applauding reason and empiricism, and denouncing dogma and the supernatural. This philosophy is typically associated with atheists, secular humanists, materialists, and skeptics who sometimes argue among themselves about what they should call their collective self. (In 2003, some advocated that they adopt the label *brights*.) Among free thinkers, consensus on such issues is not reached easily. This community—long practiced in online conflict with creationists, cranks, and scammers—fell on itself with savage severity in 2012.

Rebecca Watson is the founder of the *Skepchick* blog and a regular on the popular *Skeptic's Guide to the Universe* podcast. After participating on a panel in which she discussed behavior that makes some women feel unwelcome at conferences, including being propositioned sexually, a man joined her in an elevator at 4 a.m. and invited her to his room for coffee. She later posted a video blog about the conference, mentioned that the elevator incident made her feel uncomfortable, and urged, "Guys, don't do that."[55] Some thought that she overreacted. After the evolutionist and atheist Richard Dawkins sarcastically compared Watson's plight to a Muslim woman's genital mutilation, Watson said that she would no longer buy or endorse Dawkins's books. Others blamed Watson for exaggerating the extent of the problem at conferences and for scaring women away. Watson also became the target of misogynistic attacks, including offensive and threatening images. But Watson was supported by many bloggers on the Free Thought Blog (FTB) network, a collection of blogs that is associated with PZ Meyers, of the popular blog *Pharyngula*, where many align with social justice issues like countering sexism and racism. Eventually, claims and counterclaims of bullying were made. This case exhibited the pattern of reciprocal accusations, partisanism, rights rhetoric, doxing, and lists of bullies and misogynists.

This pattern is common. It can be found at Wikipedia and was also present in the RaceFail '09 ("writing for the other") incident mentioned earlier. The experience can be so confusing that some people simply throw in with their friends because it is otherwise such a mess. The fanlore wiki noted that even the trigger to RaceFail is a source of controversy and "several people produced roadmaps to RaceFail '09 to try to sort out the confusion."[56] In addition to these commonalities, the FTB case has two other features worth noting.

First, much like the Kathy Sierra incident in which some of the *Mean Kids* subsequently recoiled from the extent to which people could be mean, the free thought community prides itself on its harshness. Richard Dawkins is famous for the acerbity of his critique. Neil deGrasse Tyson, the affable astronomer and science communicator, once told Dawkins that he was taken aback by Dawkins's sharpness and asked if it best served the purpose of increasing the public's understanding of science. Dawkins replied, "I gratefully accept the rebuke" and likened his philosophy to something an editor of the *New Scientist* once said: "Science is interesting and if you don't agree you can fuck off."[57] Similarly, the community of bloggers and online commentators associated with Meyers's *Pharyngula* are harsh and are referred to as "the commentariat." While this term has been used to describe mainstream news pundits, in the online context it evokes a seething mass of vicious typists, much like the character in an *XKCD* comic. In "Duty Calls" a stick figure refuses to leave the computer for bed because "This is important... someone is wrong on the Internet." Meyers has written that he approves of the ferocity of his commenters, "even when they're turning their teeth on me."[58]

As one fan of *Pharyngula*'s commentariat wrote, these commenters are unpleasant but necessary:

You are cruel, rude, and unforgiving. In fact, I've found nicer people on fucking YouTube. I do not comment on *Pharyngula* that often because of you; I'm terrified of it. The thought of wading into those shark-infested waters is an absolutely horrid thought ... almost bad enough to create a phobia ... a *Pharyngaphobia*, if you will. So I think all of you owe me just one thing. Don't. Ever. Stop. Seriously. You have been one of the best online sources of skepticism that I have ever seen. The extreme intolerance for bullshit is refreshing and wonderful, even if my sensible, peace-loving, chicken-shit self is sometimes terrified of it.[59]

In addition to the irony of combative communities devouring themselves, the FTB bully-battle also found a new medium in which to express itself: Twitter "bashtags," the disruptive appropriation of an opponent's hashtag. In 2013, as a parody of the Website Who Needs Feminism, masculinists started tweeting under #INeedMasculismBecause. This was then appropriated by feminists with sarcastic tweets like "#INeedMasculismBecause the way I react to the mocking use of this hashtag shows I think mockery is the same as systematic injustice."[60] The masculinists then appropriated the tag #tellafeministThankYou. (Bashtags are especially confusing because they are hypotextual and comments can be ephemeral, prompting some to document them via screengrabs, which themselves can be faked.) In the case of the free thinkers, some FTB bloggers listed other skeptic bloggers as offensive and worthy of blocking. Those targeted called this censorship and documented this "suppression of dissent" under the #FTBullies hashtag. This tag was then appropriated, as one FTB blogger sarcastically put it: "Like the bullies we are, we FTB bloggers (and our friends and readers) crashed a hashtag originally started by a few people who just absolutely hate that we talk about social justice so much and dare to moderate the comments sections on those posts. Hilarity ensued."[61] Hash crashing and bashtags exemplify the loss of context and place in online comments today and adds a new layer of confusion to bully-battles.

Tropes and Taking a Stand

A common response by people who are hurt in flame wars, trolling, and bully-battles is to pull back, move on, or disappear. Kathy Sierra abandoned her public activity after the Mean Kids incident and has not updated her blog, *Creating Passionate Users*, since 2007. It took over five years before she ventured to begin a new blog in the summer of 2013. The science fiction author Elizabeth Bear attempted to call a cease fire during RaceFail '09 by addressing an Internet that seemingly had gone crazy. She wrote that the controversy "keeps following me home, and I'm really getting sick of it, because it's not about communication. It's about us versus them. And the problem is—the problem I see, and the reason I've been refusing to comment—is that there is no us and there is no them in this fight. It's a false dichotomy, and worse, it's a waste of energy."[62] (One of her opponents wrote that this was yet another example of white privilege because Bear could move on but "I couldn't just decide not to have a conversation about race anymore, because it follows me home. My race issues ARE my home."[63]) Adria Richards was once an active blogger, tweeter, and podcaster but went silent for about a year after she was attacked online and fired. Rebecca Watson, who was married on stage at one of the skeptic movement's biggest conferences in 2009, wrote that she would not attend the 2012 conference. Too many in the community had told her "I'm a whore, a slut, a bitch, a prude, a dyke, a cunt, a twat, told I should watch my back at conferences."[64]

Another casualty of the Free Thinkers Blog bully-battle, Jen McCreight of the *Blag Hag* blog, wrote that she was doing less blogging because people were harassing her family members and threatening to get her fired. Furthermore, "If I block people who are twisting my words or sending verbal abuse, I receive an even larger wave of nonsensical hate about how I'm a slut, prude, feminazi, retard, bitch, cunt who hates freedom of speech (because the Constitution forces me to listen to people on Twitter)."[65] For someone to withdraw from such attention is to be expected after such a harrowing experience. And after such an incident, it can take a long time for the community to recover, if ever. The cleft between those who affiliate with social justice issues (including critiquing sexism and racism) and those who do not persists in many of the online communities I've mentioned.

Occasionally, there are small victories against hater culture. In the summer of 2012, media critic and *Feminist Frequency* blogger Anita Sarkeesian proposed a $6,000 Kickstarter project entitled "Tropes vs Women in Video Games." With the money she collected, she planned to expand her earlier video analyses of stereotypically demeaning portrayals of women.

Given that sectors of the gaming industry are dominated by young men, she would not have lacked for material. Perhaps because of this aspect of the industry, Sarkeesian was also beset by a "torrent of misogyny and hate." Her YouTube videos were populated with nasty comments and flagged as terroristic so that they would be banned. Her Wikipedia page was vandalized, and she was repeatedly threatened: "These messages and comments have included everything from the typical sandwich and kitchen 'jokes' to threats of violence, death, sexual assault and rape." Insults were misogynistic and anti-Semitic, and Sarkeesian was labeled a "feminazi." (This prompted me to propose a corollary to Godwin's law: "As an online discussion about sexism continues, the probability of a woman who speaks out being called a feminazi approaches 1.") Also, a campaign was organized to have her project banned and defunded from Kickstarter.[66] In one bizarre bit of thinking, one maligner decided to balance a perceived gender inequity: "There's been a disgusting large imbalance of women who get beaten up in games. Let's add a lady to help balance things." Users could click on an image of Sarkeesian to bruise and bloody her face. The author of the game justified this as an attempt at engagement: "She has refused to address any form of criticism whatsoever, and hides behind the fact she has a vagina, claiming it's sexist to criticize her in any way. She claims to want equality: Well, here it is."[67] Sarkeesian characterized this as but one instance of "image based harassment and visual misogyny" that she encountered, which also included "everything from vulgar photo manipulation" to "creating pornographic or degrading drawings of rape and sexual assault." It would be understandable to recoil from such images, but she shared them to document what some people experience online.[68]

Despite the attacks or perhaps to counter them, Sarkeesian's Kickstarter proposal garnered almost seven thousand supporters who pledged over $150,000. In March 2013, she released her first analysis on YouTube, which has had over a million views: "This video explores how the Damsel in Distress became one of the most widely used gendered clichés in the history of gaming and why the trope has been core to the popularization and development of the medium itself."[69] Despite this success, Sarkeesian continues to receive hateful and harassing messages, as do many others online. Hate and harassment are a part of online comment for which there is no easy solution. Yet the success of her project suggests that there is an option between feeding and ignoring the trolls and haters: supporting those being attacked. I am not suggesting that the targets of abuse should engage with the trolls or become lone vigilantes. Nor do I advocate for bully-battles. Collectively, we should not ignore the trolls. Just as we should not ignore comments but curate them, we should iden- tify abusive behavior as odious and unwelcome and support targets of abuse—whether emotionally, financially, or legally.

6

Shaped: "Aw Shit, I Have to Update My Twitter"

Twitter and Facebook and MySpace; all that stuff makes you warped. We've all basically given ourselves data entry jobs. I've actually heard people say things like, "Aw shit, I have to update my Twitter." Really? You have to? That's a big priority for you?

—Louis C.K., *Vanity Fair*

"Hi, this is Jamey from Buffalo, New York, and I'm just here to tell you that it does get better." This is how a fourteen-year-old boy began his 2011 contribution to the It Gets Better project. The year before, columnist Dan Savage and his husband, Terry Miller, began the campaign as a response to stories of LGBQT youth ending their lives after experiencing harassment. Savage and Miller posted a video in which they spoke of their own difficult times as teenagers. They assured viewers that life will get better but that teens have to stay strong, make their way through high school, and "live your life so that you're around for it to get amazing."[1] Jamey took comfort in this message as well as Lady Gaga's song "Born This Way." Like Savage, Miller, and thousands of others, Jamey spoke of the difficulties he faced. He noted that at school, he was cornered in the hallways, called "faggot," and "felt like I could never escape." Like many of his peers, he also created a Formspring account, something "I shouldn't have done" because of the harassing comments that followed.

As early as the 1990s AOL had an "ask me anything" (AMA) chatroom, but it was in 2009 that this question and answer (Q&A) format became especially popular at Reddit and Formspring. At the social news site Reddit, people volunteer to provide candid answers to questions from commenters. Over four thousand Reddit AMAs have at least a hundred comments each, and the most popular ones tend to be from interesting but ordinary people (such as an emergency room nurse) and celebrities (including Weird Al Yankovic and President Barack Obama). At Formspring, young people articulate their developing identities by constructing profiles and asking questions of one another. For instance, "If you signed up for the one-way Mars mission, what would you take?" and "What do you think of Rihanna's new tattoo? Do you want a tattoo?" The questions can be embarrassing, similar to the "Truth or Dare?" game that has long been popular among adolescents. Comments can also be made anonymously; this permits a giddy sense of freedom but it can also be abused. Facebook's 2007 Honesty Box app was similarly popular and problematic, and Formspring competitor Ask.fm had been linked to cyberbullying among adolescents. danah boyd, a researcher of young people's use of social media, informally characterized this as "harassment by Q&A," noting that "While teens have always asked each other crass and mean-spirited questions, this has become so pervasive on Formspring so as to define what participation there means."[2] Like the lists on Goodreads, which can be used to censure as well as recommend, the personal Q&A format can be considered to be a *drama genre* of comment. (Informational Q&A sites, where people ask about how to do something, like remove a splinter, tend to be more staid than the sites where people ask questions of each other.)

For Jamey from Buffalo, Formspring was a place where "people would just constantly send me hate, telling me that gay people would go to hell." (As is shown in chapter 5, adults' concerns about and appropriation of the term *bullying* has led younger people to prefer the terms *drama* and *hate*.) Yet in his video, Jamey assured his peers that his life had been difficult but improved after he came out publicly about his sexuality: "I got so much support from my friends and it made me feel so secure." If family members and friends are not supportive, he recommended that his peers follow Gaga's advice: "You were born this way, now all you have to do is hold your head up and you'll go far because that's all you have to do is just love yourself and you're set." As a counterpoint to the harassment that he received on Formspring, he noted, "I have so much support from people I don't even know online. I know that sounds creepy but they are so nice and caring and they don't ever want me to die and it's just so much support for me. Just listen here, it gets better."[3] The positive power of online comment is that it allows people to connect with and support another, even if briefly and from afar.

What is heartbreaking about this video, is that Jamey's exhortations that it "gets better" were as much for himself as for others. Although he had found some support, he was not yet out of high school or free from harassment, and he killed himself a few months later. Even today, the video continues to receive comments, ranging from the supportive "Stop all the hate. RIP Jamey" to the cruel "GO KILL YOURSELF ... ... oh wait lol."

Displays of bigotry and hate are upsetting and alienating, and so we are advised to "avoid the comments." But our relationship to online comment is more complicated than "Comments are bad. Avoid." In boyd's recent book *It's Complicated: The Social Lives of Networked Teens*, she notes that many who were outraged by the bullying on Formspring did not realize that if a teen ignored a question, it never would be posted publicly. Young people were choosing to respond to cruel questions. Additionally, when boyd asked representatives of Formspring about this, she was told that sometimes an answer immediately followed a question and both came from the same Internet address. boyd concluded that some teens appeared to post mean questions to themselves as a type of "digital self-harm," perhaps to attract attention or support. In a related survey, 10 percent of respondents reported having engaged in this type of behavior.[4] For some, online comment is not simply an external annoyance that is easily dismissed. Just as giving and receiving feedback can entail significant emotion work, online comment can be an important part of the emotionally laden construction and understanding of people's social selves. For some teens, not having a Facebook page or enduring the rite of Q&A could be seen as being weird or antisocial. It is complicated. We live in a world in which we can share much about ourselves, but this is not a simple broadcast. In sharing, we also craft a sense of ourselves that is subject to the comment and approval of others.

Online comment is reactive and short, and these qualities affect people in a couple of ways. First, reactions to things (such as a comment, an answer to a question, or the liking of a photo) have come to define how people see themselves. And others' reactions to our reactions (for example, by retweeting them) are seen as a valuation of that self. Second, comment's shortness and ubiquity mean that attention is easily and often drawn online. Additionally, the fact that all of this writing can be counted and tracked, again, affects how we value ourselves and each other. Could this be making people "warped," as Louis C.K. posits at the head of this chapter?[5] How does the nonstop stream of our own and others' pictures and status updates affect self-esteem and well-being? Do the short and asynchronous bursts of comment that are processed throughout the day affect the ability to concentrate? Is the pervasive rating and ranking of people dehumanizing? These are complex questions without easy answers, but they are worth considering.

Self-Esteem, the Social Self, and Selfies

After the negative press about bullying on Formspring but before its dissolution in mid-2013, the site attempted to better police abusive behavior and counseled its users that "the hidden identity feature should never be used to ask questions that are mean or hurtful."[6] But the practice persists, if not at Formspring, then elsewhere. Surprisingly, some teens ask questions that seem bound to prompt insults. An "Am I ugly?" video from a young woman I'll call "Sarah" attracted a range of terse answers, including "ugly," "pretty," and "beautiful." Some comments are descriptive and implicitly racist: "you are black" and "you have big lips." Others are explicitly and offensively so: "your not only ugly but your black too wtf nasty." The response "I got a boner" is ostensibly a compliment but is likely intended as a vulgar insult. This video, and others like it, bear out YouTube's reputation as a repository of some of the Web's worse comments. The site Stupid YouTube Comments seeks to preserve such comments "so that people a hundred years from now can look back and take solace in the fact that the authors of these stupid comments have all since died."[7]

Other comments on Sarah's video were affirming, noting that "beauty comes from within" and "beauty is uniqueness, so you have your own beauty like everyone else." I suspect that some adolescents hear much harshness and little affirmation from their peers and risk asking questions online in hopes of hearing supportive words. But they may not initially appreciate how negative comments can be. Many commenters exhorted Sarah to remove the video. She was urged to "take this down" because strangers' opinions shouldn't matter: "Take down this video and work on your self confidence," and "This is just going to get haters and sympathy; you don't need that, find a meaning, even though you are not ugly but still you can't believe that outer beauty is everything. TAKE IT DOWN."

For a more balanced take on "am I ugly?" one can look to Reddit. Its topic (or *subreddit*) has guidelines that comments should be constructive rather than overly harsh, creepy, or insulting. Consequently, in addition to some cutting comments one can find compliments, assurances that things will get better and that personality counts, as well as suggestions for posture, exercise, and style (e.g., "you'd look good with Jake Gyllenhaal's haircut"). Submitters even express their appreciation: "Thank you all so freakin' much for the support!!!! I cried reading these comments and my self esteem improved tenfold."

Inherent in this phenomenon of "am I ugly?" posts is the notion of self-esteem. Yet it is not a simple idea. Confusion about how we conceive of self-esteem likely contributes to both its seeming scarcity (among the insecure) and overabundance (among narcissists).

Among scholars, *self-esteem* is understood as the self-evaluation of one's worth; this is part of *self-concept*, the totality of thoughts and feelings about one's self. Aside from the lonely hermit, a sense of self is understandably influenced by social interactions. That is, our sense of self is related to how we present ourselves to others and how we are received. A seminal articulation of this idea was Erving Goffman's 1959 *The Presentation of Self in Everyday Life*, in which he likened people to actors who perform on a stage.[8] At the "back stage," we are free of the constraining expectations of an audience; on the "front stage," we actively manage the impressions that others have of us.

Many are surprised and discomforted to see their own "performance." Some renowned actors, such as Johnny Depp, even avoid watching themselves on screen. Researchers have long used this discomfort to study self-esteem by exposing people to mirrors or recordings of themselves. More recently, researchers Amy Gonzales and Jeffrey Hancock compared people's exposure to a mirror and to Facebook. Traditional theory predicts that seeing one's Facebook profile would heighten self-awareness and diminish positive feelings and self-esteem, as often happens in the presence of a mirror. However, an alternative theory predicts that viewing one's Facebook profile enhances self-esteem because people can selectively control what is presented there. In their experiment, sixty-three students were split into three groups and given a survey that included ten standard questions that are used to measure self-esteem. For example, participants were asked if they (strongly) agree or disagree that "I feel that I have a number of good qualities." Members of one group were placed in a room that included a mirror. Those in the Facebook group were asked to open their profile on a computer and were left unattended for three minutes with no specific instructions before being given the survey. Members of the control group completed the survey without the presence of a mirror or Facebook. Results confirmed that exposure to their own Facebook profiles enhanced participants' self-esteem, more so if they stayed on their profile rather than browsing elsewhere and especially if they edited their profile during the experiment.[9] I think that this reaction is related to the ability to control self-presentation.

The droll *Urban Dictionary* defines *selfie* as "a picture taken of yourself that is planned to be uploaded to Facebook, Myspace or any other sort of social networking website. You can usually see the person's arm holding out the camera in which case you can clearly tell that this person does not have any friends to take pictures of them so they resort to Myspace to find internet friends and post pictures of themselves, taken by themselves."[10] Although this derisive definition is from 2009 (and the word began to circulate a few years prior to that), it was featured as the site's word of the day in 2012 and has since been widely covered in the news. The *Oxford Dictionary* named it 2013's "International Word of the Year." Despite those who willingly take amusing or embarrassing selfies, the common belief is that those who post them find them flattering. This conceit and some people's propensity for selfies (a couple a day) is an object of derision for others. Such selfies are tagged on photo-sharing sites and collected on blogs. One such blog that was popular in 2013 was *Cop Selfies*, which collected unsettling arm's-length pictures of uniformed police officers. Some *fail* selfies go viral, such as the picture of a young woman smiling up into her camera, standing in her bathroom in her underwear with the unflushed contents of her toilet visible behind her.

Most people have cringed at seeing unflattering pictures of themselves in someone else's album. We are not in perfect control of our presentation. (In online albums, sometimes the only recourse is to *untag* one's name from an image.) Some even claim that this online hall of mirrors is driving the current craze for chin augmentation, the fastest-growing cosmetic surgery in the United States. Consider Triana Lavey, a frequent user of social media, who remarked that "I have been self-conscious about my chin, and it's all stemming from these Facebook photos.... I think that social media has really changed so much about how we look at ourselves and judge ourselves. Ten years ago, I don't think I even noticed that I had a weak chin." Although people can select (or augment) their own profile pictures, they have less control over how they appear in others' photos, so some resort to altering their bodies: "Here is a weak-chin photo that I didn't untag myself in ... because I was working out really hard that summer, and I am pleased with everything else in the photo. But it's my darn chin that bugs the living daylights out of me in this photo.... You keep looking and looking, and now it's the first thing I look for in a photo. It all started with Facebook."[11]

In the 1970s, literary critic Lionel Trilling argued that the twentieth century had seen a moral shift toward personal authenticity, which displaced an earlier focus on character and sincerity. Carl Elliot's 2004 *Better Than Well: American Medicine Meets the American Dream* argued that this production of an authentic self was increasingly facilitated by pharmaceutical and surgical interventions. Elliott wrote that for many people, "it is now very important (in a way that it was not so important two hundred years ago) that others recognize and respect them for who they are."[12] But "who they are" is not taken as a given. People now have a proliferation of identities to choose from (body builder) and a multitude of ways to achieve their chosen identities (fitness regimes, diets, drugs, surgeries, and edits to an online profile). As noted earlier, however, the *paradox of choice* is that more choice does not necessarily make people happier.

The mirror is more than just a metaphor for understanding how people are shaped by comment online. It actually has been used in experiments on self-esteem. Even so, it is a powerful metaphor because it speaks to how people look to an external frame for a sense of themselves. Instead of mirrored glass, we see a reflection of our own edits and the comments of others.

Flattery versus Feedback

An unexamined premise of the discussion about self-esteem is that it is a good thing. This assumption arose in the latter half of the twentieth century for good reasons that are worth a brief historical digression, even if this assumption has led to problems today.

In 1968, following the assassination of Martin Luther King Jr., Jane Elliott wondered how to make the issues of racism and discrimination understandable to her third-graders, white children growing up in a small town in Iowa. She asked her students to try an exercise in which they would judge people by the color of their eyes. News of this experiment spread, and in 1970, journalist William Peters filmed Elliott in her classroom for an ABC television program. In the footage, Elliot began the exercise by deciding that blue-eyed people would "be on top the first day." She told her students about the many virtues of blue-eyed people and faults of brown-eyed people. She asked the blue-eyed children to place a fabric collar on their brown-eyed peers so that they could be identified from a distance. Blue-eyed children were granted perks such as extra time on the playground. They were encouraged to ignore the "brown eyes," who were not allowed to use the same water fountain as the blue-eyed children. Some students quickly divided into their respective groups, and schoolyard friends were throwing insults and punches on the playground. On the next day, Elliott reversed herself and told her students that brown-eyed children were superior. The divisions that manifested on the day earlier continued, but now the brown-eyed children had an opportunity for payback. At the end of the exercise, the children reflected on the experience and gladly discarded the collars and the discrimination that they represented. Besides demonstrating how parochial humans can be, Elliott also noted that the children's academic performance was affected: "almost without exception, the students' scores go up on the day they're on the top, down the day they're on the bottom."[13] This observation anticipated two related issues in the social sciences: stereotype threat and self-esteem.

Stereotype threat is the risk of confirming a negative stereotype: people do poorly when associated with a poorly performing group. Although the initial work on this topic compared black and white participants' performance on standardized tests, the finding is robust: hundreds of studies have found impaired performance when a negative stereotype is made salient to the performer.[14] Similarly, performance can be improved via *stereotype enhancement*, highlighting a positive stereotype about a person's group. Expectations, including those rooted in stereotypes, can significantly affect performance.

A more questionable implication that could be drawn from these research findings is that people must never communicate negative feedback: if self-esteem is positive self-regard, wouldn't flattery promote self-esteem and performance? In 1986, coinciding with renewed attention to Elliott's experiment via PBS's show *Frontline*, California established a task force to investigate the relationship between self-esteem and the epidemic of "social ills." The task force concluded in a lengthy report that there was a significant correlation between self-esteem and "how well or how poorly an individual functions in society." Consequently, "documenting this correlation and discovering effective means of promoting self-esteem might very well help to reduce the enormous cost in human suffering and the expenditure of billions in tax dollars caused by such problems as alcohol and drug abuse, crime, and child abuse."[15]

In the new millennium, the pendulum has swung once again. In a 2004 article in *Scientific American*, a group of researchers argued that the need to boost positive self-esteem in young people was a "myth" in need of "exploding." They reported that decades of research had found that self-esteem is only weakly predictive of subsequent academic achievement. Even when there is moderate evidence of correlation, there is little evidence for direct causation; that is, perhaps people have self-esteem because they do well rather than vice-versa. Despite the good intentions of some parents and teachers, artificially boosting self-esteem may lower subsequent performance. Additionally, those with high self-esteem may be more likely to engage in risky behavior, have more sexual partners, and exhibit aggression.[16]

What is critical is how adults promote self-esteem in young people. Research on learning indicates that there are both "perils and promises of praise." The *type* of praise that is given is important: does it reinforce a belief that intelligence is a fixed trait and not susceptible to improvement or that improvements can be had with further effort? This insight is based on work from psychologist Carol Dweck and her colleagues. In one experiment, fifth graders were praised for their effort or intelligence. Most of the children praised for hard work reported that they preferred "problems that I'll learn a lot from, even if I won't look so smart." Conversely, many of those praised for their intelligence preferred tasks that "weren't too hard" and "pretty easy" so that they could continue to show "they were smart." In later tests, those praised for intelligence displayed less persistence, resilience, and enjoyment and actually did worse than the children praised for effort. They were also much more concerned about their standing relative to others. These children preferred learning about other children's performance to learning new strategies and were more likely to lie to (distant and anonymous) children about their own performance.[17]

Some even argue that the excesses of the self-esteem movement have created a new wave of social ills. In their 2009 book *NurtureShock: New Thinking about Children*, Po Bronson and Ashley Merryman argued that the flattering of children has backfired. This inspired George Will, the conservative columnist, to lampoon a society in which children jump rope without ropes and soccer teams that no longer keep score.[18] Amy Chua, a professor at Yale Law School, argued that Chinese culture (and "tiger mothers") were "superior" on this point. She bragged that her children had never had a sleepover or play date, did not choose their own extracurricular activities, had to play either the piano or violin, and were expected to be the best student in every subject, except gym and drama. Chua wrote that "Western parents are extremely anxious about their children's self-esteem" and fear damaging their kids' psyches, whereas Chinese parents "assume strength, not fragility, and as a result they behave very differently."[19]

Even if one finds Chua's polemic problematic, researchers have found differences in the ways that parents approach the performance of their children. In a study that preceded Chua's book, researchers observed the responses of Chinese and American mothers toward their fourth and fifth graders' test performances. In a break between tests, children were reunited with their mothers. American mothers spent the time talking about something else, but Chinese mothers were more likely to say, "You didn't concentrate when doing it" and "Let's look over your test." The Chinese mothers did not speak harshly, were no more likely to frown or raise their voices, and smiled and hugged their children as much as the American mothers. However, the Chinese kids' scores on the second test jumped 33 percent, more than twice the gain of the Americans.[20]

At the heart of this cultural debate is our notion of what it means to promote self-esteem. If self-esteem is understood as simply thinking well of oneself and is promoted through the flattery of fixed traits (such as "You are so pretty" or "You are so smart"), such comments can backfire.

However, self-esteem can also be understood as the ability to consider and make use of positive and negative feedback (for example, about what works well and what can be improved). In psychologists Richard Bednar and Scott Peterson's book about the treatment of low self-esteem, they note that everyone receives negative feedback throughout life, much of which is probably valid. (Everyone also receives lots of positive feedback.) The extent to which we develop healthy self-esteem is not related to the number of positive messages that are received but the ways that people respond to negative feedback. Although *avoidance* is "based on psychological processes that deny or distort unpleasant psychological realities," *coping* is the ability to face threatening situations realistically and requires "personal introspection, personal honesty, and a willingness to acknowledge openly the imperfections in the self." This can be distressing, but learning how to tolerate the distress and grow from it is an important life skill.[21] In this sense, self-esteem is understood as a positive self-regard because it allows people to know how to manage and use feedback. This type of skill seems especially important given the ubiquity of comment, but we've only begun to think about how to maintain such self-esteem in the online realm. Instead, many seem preoccupied with refashioning themselves (be it at the gym or in Photoshop) and with their standing relative to others, which might actually be making them feel worse.

Social Comparison and Well-Being

In addition to chin surgery in the United States, eyelid surgery in South Korea might also give a hint to how sense of self and well-being are affected in the age of the Web. In their mirror and Facebook study, Gonzalez and Hancock found that people who browsed beyond their own profiles reported slightly lower self-esteem than those who stayed on their profiles. Just as presentation is social, so are people's expectations for themselves: we compare ourselves to others. The newly affluent South Koreans are the most wired people on the planet, for example, and also are big consumers of cosmetic surgery. One in five women are reported to have had a procedure, the most popular of which is blepharoplasty in which the eyelid is given a fold—which is the Caucasian norm but also occurs naturally among some Asians. A possible anxiety underlying this phenomenon is suggested in a popular song by the K-pop band 2NE1:

"I think I'm ugly and nobody wants to love me. Just like her I wanna be pretty, I wanna be pretty." Perhaps media exposure and cosmetic surgery are independent effects of a newly affluent society, but in a media-saturated environment, people find it difficult not to compare themselves to others ("just like her").[22]

The logic of comparison and competition in a global market is described in Robert Frank and Philip Cook's 1996 *The Winner-Take-All Society.* Their lengthy subtitle captures their thesis: *How More and More Americans Compete for Ever Fewer and Bigger Prizes, Encouraging Economic Waste, Income Inequality, and an Impoverished Cultural Life.* This competition and comparison likely diminishes our sense of well-being. Frank and Cook touch only briefly on matters of appearance, but they quote a character from Kurt Vonnegut's novel *Bluebeard* who notes that "a moderately gifted person who would've been a community treasure years ago has to give up ... since modern communications has put him or her into daily competition with nothing but the world's champions." Similarly, a hundred years ago people might have been grateful to be modestly attractive, with good teeth and hair. Today people are competing with celebrities, who themselves undergo surgery, which further shifts the standards of "normal appearance."[23] This privileging of constructed identity, proliferation of identities to choose from and ways to achieve them, and exposure to world-class successes and beauties might lead some to feel rather lacking.

This assessment of our own standing relative to others (*social comparison*) might be why Facebook use can be related to feelings of dissatisfaction, including envy (much as the wicked queen felt toward Snow White when she looked into her "mirror, mirror, on the wall"). Several studies of Facebook users have found that greater use of the site is associated with an increased belief that others are happier and doing better; the more Facebook "friends" people have, the more they feel this way.[24] It appears that it is the *passive* consumption of *others'* pages that is associated with life dissatisfaction "as it triggers upward social comparison and invidious emotions." The acronym *FOMO* (fear of missing out) is used online to describe this impression that other people are doing more and having greater fun. Holiday pictures (online and off) are especially (even if unconsciously) chafing because it is one of the few ways in which people can brag about how great their lives are without being accused of doing so. As a result, some users react to others' displays of awesomeness with more self-promotional content of their own, leading to a "spiral" of self-promotion and envy on an already "envy-ridden" Website.[25] However, many studies find correlations between these behaviors and not causations. Maybe people feel sad before they log on to Facebook? To address this, a different team of researchers sent text messages to subjects five times a day with a link to a short survey. Through this "experience sampling," they were able to draw a closer connection between Facebook use (rather than prior mood) and a small but significant decline in how people feel.[26] Even so, it is important to acknowledge that people do different things online, including on Facebook. Staying in touch with a family member from afar likely affects one differently than seeing how much fun dozens of casual acquaintances appear to be having.

Even beyond seeing photos of others' marvelous vacations, people can dampen others' moods because of a tendency to show only positive emotions to others. Researchers led by psychologist Alex Jordan wondered if the seventeenth-century thinker Charles de Montesquieu was right when he wrote, "If we only wanted to be happy it would be easy; but we want to be happier than other people, which is almost always difficult, since we think them happier than they are." He was right. In studies of people's estimations of others' feelings, Jordan and his coauthors found that people typically overestimate others' positive emotions because of "the unobservability of others' solitary experiences" and "preferential suppression of negative emotion." That is, people tend to be more cheerful in the presence of others, and even when they are not happy, they hide it: "misery has more company than people think."[27] To return to Goffman's metaphor of the stage, we are always in the presence of and performing for others in the age of ubiquitous comment.

Of course, the relationship between mood and presentation is not a simple one. As the philosopher and psychologist William James is said to have remarked, "I don't sing because I'm happy. I'm happy because I sing." Sometimes people can affect their own moods, which can be socially contagious, but the chronic suppression of emotion is associated with life dissatisfaction and depressive symptoms.[28] And shouldn't people be able to share their troubles with good friends? In a study of college students, researchers Junghyun Kim and Jong-Eun Roselyn Lee attempted to untangle the relationships between number of friends, honesty in

self-presentation, perceived social support, and well-being. When participants were asked how much social support they felt they had (such as
remembered birthdays, congratulations, and condolences), those with few
friends and those with many friends felt worse. That is, seeing only a few
friends online might be disappointing, seeing many might make one feel
affirmed and connected, but too many friends might be a sign of desperation. In fact, related work has found that those who are low in self-esteem
tended to have more Facebook friends than others, perhaps as a compensation. Also, although positive self-presentation related to subjective well-
being, so did honest self-presentation: a sense of well-being could come
from putting on a happy face for friends, but so might the social support
received after being honest about a difficult time.[29] The consequences of
technology use can vary and are dependent upon who uses it and how
they do so, but it is clear that exposure to a steady stream of others' comments can affect us.

Media Multitasking and Attention

Much of the research noted so far has been done with young people: they
are still "finding themselves" and seem conversant with technology. There
is also the matter of convenience: much of what is known about human
nature is gleaned from college students because professors can easily
recruit students into studies with a few dollars or class credit. In the age
of the Web, however, when a brief comment is easily made or read on a
whim, we are all subjects in a cognitive experiment. Many people, young
and old, feel the pleasure and pain of the compulsion to "update my Twitter." As comedian Louis C.K. quips: "Really? You have to? That's a big
priority for you?" C.K. suspects that we are losing the ability to focus on
what is important and joked that if Jesus himself returned, people would
be too busy tweeting to listen: "they'll be like 'Oh my god, Jesus is right
here in front of me right now.' 'I took a twitpick of Jesus, Oh my god,
Jesus is trending right now.'"[30]

The research focus on youth might also be because of an assumption
that they are especially proficient with technology, leading them to sometimes be characterized as "digital natives." However, media scholar Siva
Vaidhyanathan claims that this notion "that all young people are tech-
savvy" is a "generational myth."[31] In the media classes he taught, a few

students might have excellent digital skills, but most had not authored a Web page using the hypertext markup language (HTML). Other researchers have noted that race, gender, and class can affect which and how youth use online services.[32] Nonetheless, young and old tend to buy into the myth that the many who make daily (if not hourly) use of social media are similar and expert in their usage. Even actual experts are prone to overconfidence. As Nobel laureate Daniel Kahneman notes, overconfidence often emerges when someone has a good story to tell regardless of what the data say. And the story that many tech users tell themselves is that they are good at swimming in the sea of comment. Because comment is short (so it hardly seems like an interruption) and reactive (prompting a desire to respond immediately), people tend to think that they are good at it, but the data suggest otherwise.

Clifford Nass—the Stanford University communication researcher who recommended that feedback is best offered in a sandwich of broad praise, followed by brief focused criticism, and finished with detailed positive comments—has been studying the question of what he and his colleagues call "media multitasking." In a 2009 study—using college students, not surprisingly—they found that heavy media multitaskers were more likely to perform worse on task-switching tests because they are more easily distracted. Those who reported higher media multitasking in their lives tended to be slower when switching between categorizing letters (vowel or consonant) and numbers (even or odd). This does not necessarily mean that multitasking undermines the ability to task-switch. It may be that those with an inability to focus gravitate to media multitasking and confuse their preference for it as a skill. In any case, media multitasking is also related to a broader set of concerns. In a study of media use and multitasking among eight- to twelve-year-old girls, Nass and his colleagues found that negative social well-being was associated with media multitasking and use (both interpersonal, such as a phone call, and noninterpersonal, such as watching a video). Video use was particularly associated with negative social well-being indicators (for example, feeling less social success, not feeling normal, having more friends whom parents perceive as bad influences, and sleeping less). Conversely, face-to-face communication was strongly associated with positive social well-being. Again, correlation does not necessarily mean causation, but the authors concluded that "growth of media multitasking should be viewed with some concern."[33]

HOW TO BALANCE SOCIAL AND LIFE

In her 2010 book *Alone Together: Why We Expect More from Technology and Less from Each Other*, Sherry Turkle, technology scholar at MIT (and licensed clinical psychologist), was less hesitant in attributing feelings of loneliness, anxiety, and being overwhelmed to an engagement with media. She argued that multitasking now includes relations with others: "When someone holds a phone, it can be hard to know if you have that person's attention. A parent, partner, or child glances down and is lost to another place, often without realizing that they have taken leave." We seem to have found a way to spend time with others without being present, which seems magical at first but in time becomes a curse. And it is not just adults worrying about kids; Turkle's interviews with youth uncovered their concerns: "it is commonplace to hear children, from the age of eight through the teen years, describe the frustration of trying to get the attention of their multitasking parents."[34]

Even successful professionals like law professor Lawrence Lessig (mentioned in chapter 1), struggle with online distraction. In the acknowledgments section of a recent book Lessig thanked colleagues, friends, and family. He also thanked a computer program that disables the Internet for a period of time. He noted that "without the program Freedom (macfreedom.com), this book would not have been completed."[35] Similarly, I use a browser extension that does the same for a few hours every morning. For many, freedom can be found only away from the ubiquity of online comment.

A Narcissistic Epidemic?

The distracted attention, unhappy comparisons, and vain selfies (as well as *sexting* and reality TV) prompt some to worry about an increase in narcissism. The alarm is regularly raised in the media, with headlines explaining "Why Narcissism Defines Our Time" and asking "Is Facebook Turning Generation Y into a Bunch of Narcissists?"[36] Comment is short and easy to send, so it has led some people to produce a stream of comment about their lives that ten years ago would have looked neurotically self-obsessed. Some have linked young people's narcissism to the inattention of parents who are too busy with their own gadgets. Humorist James Napoli wrote a farcical story in which Facebook replaced the *like* button with a "love I never got from my parents" button:

"We all know what it is to check your Facebook status six, seven, eight times an hour to see if anyone has clicked 'like' to demonstrate how much your funny or insightful post has enriched their lives," a company spokesperson told this reporter. "We at Facebook understand that it's not that big a leap from this sort of desperate need for approval to realizing that you are expecting your imaginary Internet friends to fill the gaping hole of inadequacy you have felt since childhood."[37]

In many popular reports and discussion, the word *narcissism* is used in a general and pejorative manner that loosely means self-obsessed. Clinical definitions are more precise: narcissists are preoccupied with success, power, and beauty; they demand and reward attention and admiration but respond to threats to their self-esteem with rage and defiance. There is a paradox at the root of their disorder, as suggested by psychologists Carolyn Morf and Frederick Rhodewalt. Although it is inappropriate to claim that narcissists necessarily have low self-esteem, they can be understood as having a grandiose and vulnerable sense of self that they attempt to support by "continuous external self-affirmation." The "narcissistic paradox" (and the source of the consequent suffering) is that they yearn and reach for self-affirmation in ways that destroy the very relationships on which they are dependent: they are characteristically "insensitive to others' concerns and social constraints, and often take an adversarial view of others," so "their self-construction attempts often misfire."[38] Narcissism is typically assessed via a standard questionnaire called the Narcissistic Personality Inventory (NPI, or the shorter NPI-16). Like many personality dimensions (such as extraversion/introversion), psychological disorders are often partly assessed by way of questionnaires, but the definitions and assessments of some psychological disorders are controversial.

Perhaps the most prominent proponents of alarm about increases in narcissism are psychologists Jean Twenge and Keith Campbell. In their popular book *The Narcissism Epidemic: Living in the Age of Entitlement*, Twenge and Campbell report that data from thousands of college students show that narcissistic personality traits have risen as fast as obesity rates since the 1980s, especially for women, and this rise is accelerating. As evidence of the increase, they note that in 2006 one out of four college students agreed with the majority of items on the NPI. Also, nearly one out of ten Americans in their twenties (one out of sixteen for all ages) has experienced symptoms of narcissistic personality disorder (NPD). (Although their analysis of the increasing NPI scores in general student populations is compelling, it has been contested by others.) Twenge and Campbell attribute this to indulgent parenting, celebrity worship, reality TV, and the Internet (which provides "the possibility of instant fame and a 'look at me!' mentality").[39] They write that the Web displays and encourages an unhealthy self-obsession, as is exemplified by the names of popular sites (such as *My*space, *Face*book, and *You*Tube). They claim that many blogs "are vapid exercises in self-expression and attention-seeking" and that comment systems are biased, unfair, and promote conflict. They object to the notion that all opinions are equally valid because most commenters "have no earthly idea what they are talking about. They think they do— common among people with a tendency towards narcissism—but they're clueless. The comments that do say something intelligent are often lost in the mountain of ignorance." To counter this, Twenge and Campbell argue that parents should say no, make choices for their children, and send careful messages about competition and winning (that is, winning is not everything and not everyone wins all the time). They also advise parents to abstain from buying their children "I'm awesome" t-shirts.[40]

This is a rather broad polemic. Fortunately, there are studies that look at social media and narcissism, though they should be understood with some caveats. First, research on this topic tends to use NPI scores that are correlated with the behavior of users or the features of their Facebook profile. Correlation does not equal causation. Social networks might attract those with narcissistic traits, or perhaps people are more narcissistic when online. Second, even finding a correlation between narcissistic personality traits and media use does not mean that people have a clinical disorder.

With these qualifications in mind, what have researchers found? In a 2008 study, high NPI scores did correlate with higher quantities of Facebook interaction (number of friends and number of wall posts) and with the profile's attractiveness, self-promotion, and sexiness—as assessed by independent research assistants.[41] In another study, the authors noted that online social networks are an environment that is well suited to narcissism since they permit people to maintain hundreds of shallow relationships in a highly controlled environment. They found correlations between NPI-16 scores and Facebook use (number of times checked and time spent) and self-promotion via the main photo and status updates. They also discerned a gender difference in that men displayed more self-promotional content in About Me and Notes, whereas women did so in the Main Photo section.[42] In short, there is evidence that well intended efforts to "promote self-esteem" in a society that is increasingly media saturated, competitive, and global are associated with some people being more self-obsessed and dissatisfied than they would have been decades ago. I believe this is heightened by something I refer to as quantification.

The Logic of Quantification

The 1979 romantic comedy *10* made stars of Dudley Moore, playing an infatuated man in a midlife crisis, and Bo Derek, playing his fantasy woman. Derek's slow-motion jog on the beach can still be found on YouTube, and the ranking of another's attractiveness from one to ten is sometimes still called the "Bo Derek scale." (In the film, Derek's character broke the scale as an eleven.) People are predisposed to scales and rankings and imbue them with an almost magical potency. In the 1984 band mockumentary *This Is Spinal Tap*, the lead guitarist proudly shows off his amp, which goes to eleven—"one louder" than other amps. Yet he has no idea how to respond to the question, "Why not just make ten the top number and make that a little louder?" He pauses, befuddled, and responds "but these go to eleven."

In the preceding sections, I have discussed how comment today affects people's sense of self, well-being, and attention. Its ubiquity reflects (and perhaps distorts) images of ourselves and others, and its shortness and asynchronicity lead to distraction. Comment also can be quantified, as seen in the earlier chapters on reviews, ratings, and manipulation. Comment can be as short as a simple click of a button (such as a *like* or +1); these can easily be tallied and used to evaluate people in disconcerting ways.

The rate-everything craze that is exemplified by the Jotly app (discussed in chapter 3) is only one example of a proliferation of people-rating apps and sites. In 2008, PersonRatings.com launched with the intention of being a "Yelp about people" that permits anyone to opine about others. Users could leave comments without registering, with no way for the target to opt out. The site was widely criticized and ridiculed and closed within the year. In 2010, the employment Website Unvarnished.com launched to similar criticism about how members could anonymously rate one another's professional performance. Dozens of venues (including *The Economist*) published stories about the site, and most were incredulous of the concept and its success.[43] To encourage constructive reviews, Unvarnished required people to use Facebook to login, so although reviews were anonymous, the administrators presumably could use the Facebook account to identify bad actors. Additionally, people could join only by being invited by a member and reviewing that person, which would likely be positive and hopefully create a constructive culture. The site relaunched as Honestly.com in the same year, claiming that it had succeeded in creating a positive community: 65 percent of ratings were five-star, and only 2 percent were one-star.[44] In 2012, the organization changed again: both the name of the project and its philosophy of crowd-sourced reviews were dropped. Under the name TalentBin, the service created profiles of people that were based on professional information found throughout the Web. (It was purchased by the employment Website Monster.com in 2014.) KarmaFile was fairly savvy, letting colleagues rate the expertise, motivation, and professionalism of their peers.

An aggregate score was then created with an associated confidence level ("score strength"). Those reviewed had the ability to see their raters and aggregate scores but could not link a specific rating to a particular person. Those reviewed could ask the site to reject inappropriate reviews (although the applicant's rationale for the rejection would be part of the profile), or they could choose to hide their profile.[45] Despite these refinements, KarmaFile's Website and Twitter account have not been updated since early 2013.

Putting aside the failures of these particular ventures, why are people keen to rate and rank others? As is shown in chapter 1, from a communications perspective, rating can be seen as a form of gossip. It is a convenient way to assess the relative social standing of peers and form alliances. Even the rating of others' attractiveness can be seen as a form of social posturing: publicly assessing others (whether asked to do so or not) is a gesture of social power. The chapter on manipulation shows that an economist is likely to think of ratings in terms of information asymmetry. How can people best transact with others without knowing about their characteristics and abilities (that is, how to avoid a lemon)? Another way of thinking about this is from the perspective of a social theorist or historian. In 1983, sociologist George Ritzer published an article arguing that the McDonald's "rationalization" of food production can serve as a useful model for understanding changes in society. He argued that McDonald's succeeded because of things like efficiency and predictability and that these same forces now shape our lives at work, in school, and even when on holiday. "Calculability,"a drive toward quantifiable measures, is a defining characteristic of contemporary "rational" society. Why? Ritzer noted that quality is "notoriously difficult to evaluate" and yet computers are good at counting: "How do we assess the quality of a hamburger, or physician, or a student? Instead of even trying, in an increasing number of cases, a rational society seeks to develop a series of quantifiable measures that it takes as surrogates for quality."[46] Thirty years later, apps and sites can rate and rank both burgers and doctors. These quantifiable measures are then subject to manipulation. And we are just beginning to ask how these systems can perpetuate social biases. Even absent purposeful manipulation, some might be harmed in a world in which everything and everyone are rated. For instance, the car service app Uber lets the passenger rate the driver and vice versa. Might an older passenger be avoided because he received only four out of five stars from past drivers who had to place his walker in the car trunk? As the practice of rating increases, so does the leakage of our biases into the world.

Ritzer is not the only one who has noted this trend. Since the early twentieth century, dozens of thinkers have identified and decried the rationalization of people as instrumental means. What is interesting about rationalization in the twenty-first century is that it has been crowd-sourced. The assessment and evaluation of people by large institutions is now complemented by dozens of other sources, including the comments from our online "friends."

Out with Klout

Klout—whose motto is "Influence your world"—is (in)famous for making use of the information that can be found online. According to its Website, it collects over twelve billion "signals" every day from eight different social networks (like Twitter, Facebook, and Wikipedia) to compute a score representing a person's influence (President Barack Obama is a ninety-nine; Justin Bieber is a ninety-two):

> The majority of the signals used to calculate the Klout Score are derived from combinations of attributes, such as the ratio of reactions you generate compared to the amount of content you share. For example, generating 100 retweets from 10 tweets will contribute more to your Score than generating 100 retweets from 1,000 tweets. We also consider factors such as how selective the people who interact with your content are. The more a person likes and retweets in a given day, the less each of those individual interactions contributes to another person's Score. Additionally, we value the engagement you drive from unique individuals. One hundred retweets from 100 different people contribute more to your Score than do 100 retweets from a single person.[47]

Of course, as soon as something is quantified, others often are willing to manipulate it. There are hundreds of pages with suggestions for how Klout scores can be improved. Some are the "right way" (for example, tweet with @names, and connect one's social network accounts), and some are not. Klout warns that it closely monitors its data and its algorithms look for "inauthentic behaviors—spambots and the like. The Score will continue to evolve and improve as we add more networks and more signals." During a controversial 2011 algorithm adjustment, some people objected that their scores had plummeted by as much as twenty points.

Beyond bragging rights, there are other benefits to actively participating at Klout. Much like Amazon Vine reviewers, those who do well in the scoring system ("influencers") can receive "perks" and receive free or discounted products and services. The "Perks Code of Ethics" notes that no one is obliged to comment about a perk but that if you do, "Klout asks you to disclose you received a sample."[48] However, in Klout's own example, in which users received a free digital camera, the tweets raving about the camera make no such disclosure. Klout is not alone in this space; there are also Kred, PeerIndex, and Radian6, among others. At Kred, highly rated people can become "leaders." One such leader condemned Klout's perk program and lauded the higher usefulness and ethics of Kred "rewards." At Klout, "For every cool well-publicized Perk—like entry to Cathay Pacific's lounge or a weekend with a Chevy Volt—there were also plenty of examples that made it a figure of derision in the social media community, like modest hair gel samples and gifts that were out-and-out irrelevant to recipients (dog food offered to people that didn't have dogs, and so forth)."[49]

Critics of Klout include more than just its competitors. Beginning in 2011, articles began to appear online with titles like "Klout Is Bad for Your Soul" and "Delete Your Klout Profile Now!" People had begun to resent the allure of the service. Klout CEO Joe Fernandez stated in an interview that he did not originally appreciate how curious people would be to find out their scores: "I didn't think about the ego component of having a number next to your name. ... we're trained to want to grow that score."[50] Despite his disclaimer, people are preoccupied by rankings, and it is unlikely that he failed to think about this human tendency. In fact, I think he probably counted on it as evidenced by the site's automatic enrollment and scoring of users. A *New York Times* article reported on how Klout had "dragged the unwitting across the Web" by creating profiles and scores for users' Facebook friends, including their children. Klout user Maggie Leifer McGary found that her thirteen-year-old son had been scored. (When she told her son this, she said he asked "What's my score? How many points do I need to get stuff?") Klout subsequently said that it would allow users to delete profiles and would not automatically score the Facebook friends of its users.[51] Despite this assurance, it has not always been easy to find the page for profile deletion.

Other objections included resistance to quantification and concerns about who benefits. According to social media blogger Rohn Jay Miller, the advertisers, not the users, are the ones who really benefit: "Social communications should be for the benefit of the people doing the communicating. Influence cannot be measured, just as beauty and cool cannot be measured. Measuring 'social influence' tries to sell the lie that such things as 'social influence' and 'connectedness' can be measured quantitatively, then acquired, packaged and sold to the highest bidder."[52] Miller believes that Klout will succeed only to the extent that people believe in its "lie" and recommends that they delete their profile and move on. Similarly, Alexandra Samuel, another social media author, notes that the "cellular-level awareness" of retweets and +1s empower only the social network platforms: "The more we each pay attention to still-questionable metrics like Klout or Twitter mentions, and the more we choose to structure our work and lives to optimize them, the more they matter. We are creating a world in which we live our online lives as a scorecard."[53] Samuel advocates challenging this self-fulfilling tendency by way of a "Social Sanity Manifesto" that lists "commitments that you can make to escape the measurement trap, and bring some humanity to the people you interact with online." It includes deleting your Klout profile, connecting only with those you actually know, not gaming online metrics, and ignoring (or at least pacing the consumption of) ratings and analytics.

For many people, it is hard not to believe in the "lie" when running in the quantification rat race. This can prompt anxiety, as Klout user Calvin Lee noted during a holiday: "I was worried that brands couldn't get in touch with me. It's easy for them to forget about you. And I knew my Klout score would go down if I stopped tweeting for too long."[54] Or it might prompt manipulation. Data analyst Gilad Lotan noted that when he purchased fake Twitter followers, these "fake friends" had the "real benefit" of attracting additional authentic followers. Having thousands of followers made him look more credible, allowed his Klout score to shoot upward, and improve his placement in Bing's search (which collaborates with Klout). The ease and effects of this manipulation prompted Lotan to fear that "What used to be completely frowned upon, is now effectively considered an act of social media optimization."[55] There also have been reports of people who were not hired or promoted because of their scores. Even dating services have claimed to use Klout scores in their matching algorithms.

Quantified Relations

Contemporary dating exemplifies many of the themes of this chapter. People are competing in a market in which some assess others on a ten-point scale. Although many of the young men who use the Bo Derek scale today were born after her jog on the beach, the practice continues. In fact, one can find long conversations online about the scale, such as an argument about its strictness: "The biggest mistake men make with the base ten rating scale is not applying hard limits. If your scale goes from 1–10, no girl can be an 11. This is science, people, and in science everything has to fall on or within the limits of your scale." This preoccupation with quantification is especially pronounced in the self-styled "seduction community," wherein a "hot babe" (HB) is commonly understood to be a 7.5+. On a seduction "lingo" site, the entry for "Decimal Rating Scale" notes that the scale provides a shorthand for men to record and discuss their exploits: "Field reports will often read, 'got a HB8 blond school girl to go back to my place' as an indication of the level of her physical beauty."[56]

Women seem to be less interested in identifying the subtle differences between a seven and an eight and more with avoiding the "dicks." I first encountered an Internet dick list in the 1990s. Nikol Lohr, proprietor of the Website Disgruntled Housewife, described its origins in 2002: "The Dick List began 7 years ago at the Pasadena house. It was a very girly house for a long time. It was also a very listy house. So in honor of both of those characteristics, we developed an oft-revised, publicly posted Dick List on a little white board in our kitchen." The list had two purposes: "1) promoting girly solidarity through bile-spewing; and 2) reminding us that certain guys were real dicks." The online version typically obscured last names when browsing, but the database was searchable by name and location.[57] This practice continues today with apps like Lulu ("A girls-only space for insights on life and love") and sites like Woman Savers ("Research & Rate B4 U Date"). As I argue in an earlier chapter, lists can prompt much drama.

As seen at Klout, people do become preoccupied with their standing relative to others. In the studies of feedback to children, the kids who were most concerned with scores failed to learn and improve, tended to misrepresent themselves, and were preoccupied with the performance of others. In dating, there can be a similar preoccupation with one's "league" and trying to impress. I recall reading the complaints of a woman about a man's need to be rated (his car, his outfit, and the restaurant) as part of his attempts to impress, but his rating preoccupation led her to think that he was subpar.

The abundance of choice and scrutiny in dating may paradoxically make it more difficult to find long-term relationships, as is discussed in Dan Slater's widely read article "A Million First Dates." Slater notes that a buffet of choices might lessen daters' satisfaction and commitment (via the paradox of choice). As the selection grows, daters might become "cognitively overwhelmed" and more likely to make careless decisions, which leads to less compatible matches. An interviewee in the piece, a single man in his thirties, reflected on a relationship from a decade earlier: "I'm about 95 percent certain that if I'd met Rachel offline, and if I'd never done online dating, I would've married her. At that point in my life, I would've overlooked everything else and done whatever it took to make things work. Did online dating change my perception of permanence? No doubt. When I sensed the breakup coming, I was okay with it. It didn't seem like there was going to be much of a mourning period, where you stare at your wall thinking you're destined to be alone and all that. I was eager to see what else was out there."[58] But the seeming glut of many other singles can turn eagerness into anxiety. New innovations are continually deployed so that daters can filter through as many prospective mates as quickly and efficiently as possible. This began with online dating, but even for those who want a more personal touch, face-to-face contact can be brief. Speed dating—where candidates rotate through brief interactions and note the people they would like to speak with further—has been around for several years. A handful of new dating services combine online profiles, Klout scores, and speed dating into "PowerPoint parties" where each dater has six minutes to convince the assembled group of their romantic virtues.[59]

Because of a mass-market advertising campaign, it even is possible to estimate the value of relationships. In 2008, Burger King introduced its "angry whopper" to the United States ("deliciously hot with a kick"). As part of the media campaign, the company launched the Whopper Sacrifice, which awarded a free Whopper sandwich to people who downloaded the Burger King Facebook app and unfriended ten acquaintances. The ten unfriended folks were informed that they had been dropped for the sake of a sandwich that costs $3.69. At least 23,000 people participated before the campaign was discontinued.[60] (Facebook asked Burger King to alter their application, and the company was approaching its planned limit of

MONETIZE YOUR SOCIAL GRAPH

25,000 in any case.) Of course, people could easily be refriended, and this should not be interpreted as a valid finding on the worth of networked relationships. Yet it speaks to the broad recognition of the self-serving and quantified character of online life.

How to Thrive amid Comment?

Substantive research suggests that in the age of the Web, we are changing, being shaped. The distraction, ubiquity, and comparison that are inherent to online comment cannot be denied. The question is are we being "warped," as Louis C.K. suggests? It's a difficult question to answer.

Social scientists' findings based on brief and artificial interactions among college students do not necessarily generalize to the actual behavior of larger and more varied populations. And reports of correlations in larger populations do not necessarily mean causation. But people make arguments based on many reasons—research, interviews, lived experience, and history. Opposing sides contend that things are getting worse and conversely that things have never been better. Confoundingly, it is possible for both of these things to be true for different people or even for the same people at different times or in different aspects of their lives. When I awoke one morning to the sound of helicopters and smell of smoke, I checked local online news sites, which did not report on the event for thirty minutes. But people were tweeting the details of a house fire a few blocks away. Rarely does anything happen in the world that I can't wait a few minutes to learn about, but sometimes it is convenient. Nonetheless, the constant trickle of novelty and news can be a great distraction. So like Professor Lessig, I sometimes disable my connection to the Web when I need to focus.

But we need not be paralyzed by indecision about whether we are riding on the highway to heaven or going to hell in a hand basket. As the subtitle to Howard Rheingold's book *Net Smart* declares, we can learn "how to thrive online." In his chapter on attention, he notes that he begins the first day of class by asking students to turn off their phones, close their laptops, and spend one minute being mindful of how their attention shifts. Later in the semester, he allows five students to have their computers open at a time. These little tasks prompt students to think— perhaps for the first time—about their attention and use of technology. He recommends that readers who want to improve their focus can use techniques and tips, such as creating blocks of uninterrupted time with a clear intention of what they want to achieve.[61] I use his book in my own course on "Communication in the Digital Age," and I think that students appreciate it. Instead of hyping digital natives or decrying a generation of narcissists, we need to find ways to develop a robust self-esteem that can handle ubiquitous comment, an attention that is resistant to digital temptation, and relationships that are safeguarded from the colonizing logic of quantification.

7

Bemused: "WTF!"

Angle is wrong (2/5 stars): I tried the banana slicer and found it unacceptable. As shown in the picture, the slicer is curved from left to right. All of my bananas are bent the other way.

—J. Anderson, Amazon review of Hutzler 571 Banana Slicer

"Angle Was Wrong" Was Wrong (5/5 stars): I can't believe anyone could be so inept as to think that they couldn't slice their bananas because they bent "the wrong way." All that person has to do is to buy the model 571C Banana Slicer that is for bananas that bend the other way. Although I prefer left-bending bananas, I got both the 571B and the 571C so that when shopping, I don't have to have the hassle of finding bananas with the correct polarity. I hope "Angle Was Wrong" sees the light and removes that harsh one-star rating for this indispensable product duo.

—H. Madizon, Amazon review

In 2010, Ken Fisher, editor of the technology news site *Ars Technica*, posted an article about "cleaning up our comments." He wrote, "Today, I would like to discuss with our community the recent decline in both quality and civility in our front page news discussions/comments." Fisher then solicited suggestions from his readers for how to address the situation. The first comment to Fisher's thoughtful query was literally this: "First!"[1] This short exclamation is both inane and intriguing—much like "fail." For those who read comments and read them in the order posted, "first" is often the first comment, the second comment, the third comment, and so on. These are the result of competitive commenters submitting their firsts within seconds of each another—but there can be only one first. The inanity of this comment is that it contributes nothing to the discussion and is likely to be submitted before the article could be read so the commenter can claim the honor of being first. As far as honors go, it is slight, so why do people even bother?

Beyond the pleasure of seeing one's name in a prominent position, the first (or at least early) comment can be a valuable asset. I came to appreciate this over a decade ago on the social news site Slashdot. Like many sites, Slashdot permits active users to rate comments, which have a cumulative score from worst (−1) to best (+5). Readers can then filter out comments below a threshold. The system keeps the worst comments from being visible, but I noticed that some of the most informed comments were also hidden. This *rush and slash effect* privileged comments that were written within the period that most people were likely to rate them. Because those who were likely to rate comments were the most active users, they typically did this within hours of a story's posting. After that, the raters had moved on. I found that the average age of a comment with a rating of "4 or higher" (where I set my reading filter) was just over an hour and that I typically would not see any comments older than eight hours. Early comments often received more attention than they deserved.

This rush to post an (often slapdash) comment is related to a widely observed phenomenon: *preferential attachment,* or the idea that "the rich get richer." For example, when new users look for a blog or podcast to subscribe to, they are likely to select (or attach) to those that are already popular. (Because of this, the social media marketplace is characterized by firms getting users first and worrying about profits later.) This is not to say that first mover victors have no merit, but equally compelling competitors who were late to the game might suffer because of their tardiness. Similarly, in the context of comments, early posts are likely to attract a disproportionate number of responses and votes. Early comments can also bias what follows. In a 2013 study, researchers took over a hundred thousand comments during a five-month period and did one of three things to each: they rated the comment up (positive) or down (negative) or did nothing (control). Although the researchers' down votes tended to be countered and neutralized, an initial up vote increased the likelihood of a subsequent up vote by 32 percent. In some conditions (such as on political and social topics), this positive "herding" increased final ratings by 25 percent on average.[2]

Because preferential attachment fosters the dominance of early movers, we now also look to who is *trending*. Unlike a fixed number (such as two million readers), trending indicates how quickly something is gaining attention (for example, twice as many readers as last week). This, like

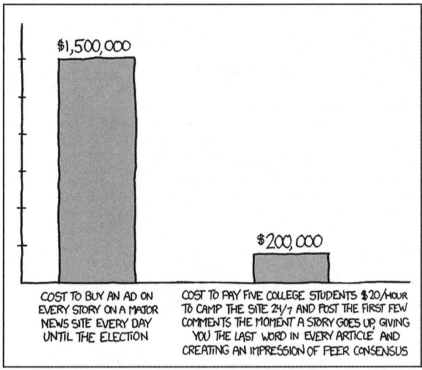

$1,500,000

$200,000

COST TO BUY AN AD ON
EVERY STORY ON A MAJOR
NEWS SITE EVERY DAY
UNTIL THE ELECTION

COST TO PAY FIVE COLLEGE STUDENTS $20/HOUR
TO CAMP THE SITE 24/7 AND POST THE FIRST FEW
COMMENTS THE MOMENT A STORY GOES UP, GIVING
YOU THE LAST WORD IN EVERY ARTICLE AND
CREATING AN IMPRESSION OF PEER CONSENSUS

THE PROBLEM WITH POSTING COMMENTS IN THE ORDER THEY'RE SUBMITTED

nearly everything, can be gamed. For example, the Web app Thunderclap allows social media campaigns to organize their supporters to post a message at the same time, "spreading an idea through Facebook and Twitter that cannot be ignored."[3]—because such a message will appear to be "trending" with a stunning growth rate.

These first-mover advantages can also be seen in the increasing dominance of Amazon Vine reviews. Select customers are given "new and pre-release items" for free if they quickly post reviews, which then have a disproportionate standing. For example, the first four most helpful reviews for a popular laser pointer are all Vine reviews that were posted within four months of the products' availability.[4] Vine reviews might be expected to be dominant for a new product, but this laser pointer has been around since 2009 and has over six hundred reviews (averaging 4.5 stars). It seems unlikely that the most helpful reviews really are those that were posted within months of being received by people who got it for free. There are obvious advantages to priming the pump.

In earlier chapters, I address how comment in the age of the Web can inform, alienate, and manipulate us. The example of "First!" shows that these messages also can be confusing and amusing: we are bemused. Online, this is often expressed via the textual exclamation "WTF?" (what the fuck?). This acronym has become a part of the lexicon through which commenters express their surprise and befuddlement. In this chapter I explore this bemusement through various examples. And while I began with those that shout "first," I will conclude with those that whisper. But, first, I consider comments that are confusing, funny, and surprisingly revealing of our biases.

Perplexing People: "Saved Our Son's Life? 4/5 Stars"

A digital bit can represent two states: true or false, 1 or 0, like or dislike. Between the single bit indicated by *like* and the verbosity of prose sits the constellation of five stars. Although we might appreciate the stars' ability to illuminate, they also can befuddle, such as a review of a carbon monoxide alarm entitled "Saved our son's life—4/5 stars." A screen capture of this review went viral, and comments at the discussion sites Reddit and Imgur revealed varied responses to online ratings, including questions of expectations and competency.

People found this 4/5 review to be a perplexing example of the inscrutableness of others. One commenter quipped that "some people are just impossible to please."[5] Is this the case? (And is it even a genuine review?) After some searching, I found that it was indeed a real review on a Canadian housewares site by user KayBe:

This morning my son called me at work to tell me the alarm was going off.... I called my husband (who luckily was not far away) and he headed home. He checked our furnace ... and discovered the burner was "carboned up" so it wasn't burning properly.... If he hadn't caught it and cleaned it the CO would have built up and our son might not have woken up in time. Scary thing to have happen but so glad this unit did it's job![6]

Based on her other reviews, KayBe was not impossible to please. In her four reviews on the site, she gave five stars to a nutritional scale and a paraffin bath, used to soak sore hands or feet. Nonetheless, her review of the carbon monoxide detector sparked parody and complaints. Some (humorously) took the four stars as a reflection of KayBe's lack of fondness for her son: "If it would have been our favorite son, 5/5 stars," and "Yeah, sure, it saved his life, but is he really that great of a kid?" Other comments sarcastically implied that some products are inherently more worthy than others: "You know, it was pretty good and saved my son's life and all, but let me tell you about my new Revlon Paraffin Bath!!! It's amazing!"[7] In reality, KayBe would likely give five stars to the idea of the gas detector as well as its performance but felt that this particular detector fell short. What might be to blame? Her four-star reviews reflected a concern about quality and durability. Some thought this was trivial compared to saving a life: "Plus—saved son's life. Minus—appearance and quality could use a little work...."[8]

These parodies likely arise from our (often confounded) assumption that others have similar expectations and competencies. In an earlier chapter I noted that in the online marketplace we lack the tangibility of a product. We wonder if the product is the right size and color, and we worry if the vendor is honest and will fulfill orders promptly. This is the information asymmetry that reviews and ratings are intended to address and it is a space in which our communications with others are paramount. Yet implicit in any communication is some tacit knowledge that informs how people make sense of the messages they send and receive. This is also true of reviews. We assume that others have similar expectations and competencies. As in the review of the carbon monoxide detector, however, sometimes there is a mismatch. Befuddled commenters wonder how someone can write that a product "works great!" but not give it five stars.

Sometimes we are afforded more insight into others' minds, and so our confusion is replaced with disbelief. In the discussion of KayBe's review, one commenter, a server in a restaurant, complained that he once gave a choking customer the Heimlich maneuver. It turned out that the diner was also a secret shopper, meaning that a perfect review from him could have earned the server a $100 cash bonus: "the asshole marked me down ONE point for not giving him the dessert menu like i was supposed to. like i'm going to ask a guy who nearly choked to death if he wants more food." Indeed, people seem quite perplexed by the unreasonable expectations of others. Parodies of such reviews include a complaint of a Mini Cooper that couldn't "transport my sectional sofa and king size bed when

I was moving. What a piece of crap car!—1/5."[9] Beyond others' confusing expectations, rating systems themselves can be mystifying.

Stupefying Systems: "How Would You Rate Your Pain on a Scale of Zero to Ten?"

As much as other people can be confusing, rating systems themselves are a puzzle, especially their scale (how many stars should there be?), applicability (what aspect of a product or service do they apply to?), and meaning (what exactly does "1/5" mean?).

With respect to scale, is five stars enough to capture the depths of the human heart? It seems to depend on what is being rated and where the rater is in the world. In North America, for instance, medical clinicians tend to use an eleven-point numerical scale for pain—to the frustration of many (myself included). Comedian Brian Regan tells about having excruciating stomach pain and being asked by hospital staff "How would you rate your pain on a scale of zero to ten?" He did not want to be outdone by other patients competing for medical attention but also hesitated to go too far. Because he had heard that a broken femur is the worst pain a person can experience, he decided that must be a ten, and "I was worried that they would've heard about me at the femur ward and hobbled into my room: 'Who the hell had the audacity to say he was at a level ten!?! You know nothing about ten; give me a sledge hammer and I'll show you what a ten is about Mr. Tummy Ache."[10] He figured that nine must be reserved for childbirth—and giving birth with a broken femur "must be hell"—so he said "eight," a decision he was happy with since he got morphine. Research has revealed that quantifying pain actually can vary according to patients' expectations (how long will the pain last?) and disposition (are they anxious or optimistic?). Since the 1970s, dozens of other pain scales have been proposed and researched with respect to their reliability and sensitivity across varied populations (children, adults, and older adults). The most popular pain scales are the numerical (zero to ten), verbal (mild and severe), and cartoon faces (crying and smiling).[11]

Social scientists are fond of a scale that was first proposed by American psychologist Rensis Likert in the 1930s. For example, "Online comment should be avoided. Do you strongly agree, agree, disagree, or strongly disagree?" Since the Likert scale was developed, much research has been done on how many points it should have. Five to nine points is thought to be a good balance. An odd number of points permits a neutral (middle) value, whereas an even number forces the respondent to take a side.[12] In addition, subjects' responses to the scales appear to be cultural. A study of participants from ethnic and mainstream supermarkets in California found that those who identified as Japanese reported more difficulty with Likert scales, those who identified as Chinese were more likely to skip questions, and on measures of positive emotions both groups were more likely to select a neutral midpoint than their (more positive seeming) American-identifying peers.[13]

People are often perplexed by the question of applicability: what do ratings apply to? Some systems have multiple dimensions. Yesteryear's Zagat had a sophisticated system; in addition to cost, it rated food, decor, and service on a thirty-point scale. As Zagat is integrated further into Google, its thirty points are being translated into five stars.[14] A single five-star variable is certainly simpler than multiple variables using thirty-points; however, it leads to confusion when people want to express concerns about things other than the product, such as shipping. As seen in review parodies, this is one of the greatest frustrations that readers have with others' reviews: "UPS delivered this broken, and Amazon is shipping me out a new one no questions asked—1/5 stars." The same complaint is applicable to app reviews: "Downloaded it, but it never showed up in my apps. This shit broken!—1/5." People who order the wrong thing can be especially irksome in their reviews: "It was exactly as described, but I realized soon after that I ordered the wrong thing. The return went smoothly—2/5."[15] This has led some to propose that ratings be made more detailed. For instance, one commenter proposed that Yelp follow Zagat and break out quality, service, and price. At Amazon, many people do not seem to realize that it is possible to "leave seller feedback" independent of the rating given to a product. In 2012, Amazon also began experimenting with the rating of product attributes, such as battery life. Even so, people still overload the star with their grievances, such as submitting negative book reviews because they experienced technical difficulties with Amazon's Kindle ereader.

Finally, there is the issue of meaning: a symbol can mean different things to different people. Today the thumbs-up sign is taken as positive, but for a Roman gladiator it might have meant death or clemency. (While the gesture endures, its meaning has been lost to history.) More recently, does a "6/10" rating mean slightly positive or a "D" and close to failing? Comedian Brian Regan might be pleased to know that there is a five-point pain scale that includes functional definitions: pain can be described as nonexistent, mild (annoying), moderate (interfering), severe (disabling), and "as bad as you can imagine." I prefer these functional labels myself, which some folks improvise. One Amazon reviewer includes the following guide with his book reviews:

Dav's Rating System:

5 stars—Loved it, and kept it on my bookshelf.
4 stars—Liked it, and gave it to a friend.
3 stars—OK, finished it and gave it to the library.
2 stars—Not good, finished it, but felt guilty and/or cheated by it.
1 star—I want my hour back! Didn't finish the book.[16]

Beyond the meaning of the stars for an individual, there is also the question of their relative and cumulative meaning. Laura Miller, a book reviewer for *Salon* and the *New Yorker*, has written that many Amazon readers are skeptical of high average ratings and develop their own approaches to reading the comments. Many people discount the five-star ratings as being written by friends of the author. Some jaded users consider only the single-star reviews. There are also "those who read in the middle ratings because they assume that only ulterior motives or sheer cussedness would provoke a reviewer to either of the extremes."[17] Others restrict themselves to the "most helpful" reviews. (As discussed in the chapter on manipulation, these are not necessarily good indicators of whether a review is fake or not.) Seasoned Amazon shoppers direct their eyes to the rating histogram and its horizontal bars. For instance, I tend to limit myself to products with lots of reviews and compare the ratio of five stars to single stars for each product. In a perfect world, I could create my own filter that eliminated products with only a few reviews as well as reviews from nonverified purchasers and from single-product reviewers. But I am a review addict, a "maximizer," and I doubt Amazon would want to complicate their interface for the likes of me.

Absurdities: "Elegant Design—Just for Her"

The comics included in this book illustrate the extent to which people are bemused by comment. (*XKCD* is widely known online and *Geek & Poke* is a personal favorite.) Product reviews themselves are another way in which collective confusion and amusement manifest themselves. In an earlier chapter, I note that many commenters joined George Takei in reviewing a drum of personal (sexual) lubricant and one unlucky fellow's farcical endorsement subsequently appeared as a Facebook ad. Sex lube is one of many products that have attracted comical reviews on Amazon. The "funniest" or "stupidest" Amazon product reviews are a common subject for lighthearted *listicles*—online articles formatted as lists. Not all such lists intersect, but a few items do and are now considered classics. For instance, Amazon offers "naturally occurring radioactive materials" for about $40. Its radioactivity is low but sufficient for testing instruments.[18] The merchant also sells Geiger counters, so this looks like a genuine product that caught the imagination of reviewers. George Takei posted a review (his standing as an Amazon Top 500 reviewer is based on such farce), but my favorite review recommends that the materials be purchased with the Oxo Good Grips Salad Spinner "so you can centrifuge it and increase its applications."[19] Amazon's page for the product notes that the ore is often purchased with Canned Unicorn Meat, hinting it is likely purchased as a gag gift. These "frequently bought together" recommendations are another source of bemusement, such as a recommendation for men's khaki pants "because you rated *Star Wars Trilogy*."[20] Other product recommendations that come by way of *Star Wars* include a twelve-cup programmable coffeemaker and a nose and ear hair groomer. Amazon's algorithms appear to consider *Star Wars* fans as older, caffeinated, men who wear business casual.

A product that appears in nearly all lists of funny Amazon reviews is the Denon AKDL1, a fancy Ethernet cable that sells for $500 (a similar cable can be bought for less than $10). Some self-identified audiophiles spend a lot of money for such equipment, which offers little appreciable improvement over much less expensive gear. This prompts heated debates between the subjectivists (those who say they can appreciate and are willing to pay for such differences) and the objectivists (those who say such products are scams that are perpetrated on deluded dilettantes). In the reviews of this cable, the objectivists had their fun—as does George Takei. For instance, one reviewer complained that the "transmission of music data at rates faster than the speed of light seemed convenient, until I realized I was hearing the music before I actually wanted to play it." Another review referenced another much parodied (dairy) product: "I accidentally dropped one end of my Denon cable into a glass of Tuscan whole milk I was drinking. Later when I finished my milk (yeah, I still drank it; should I not have done that?), my right arm (lost in an accident in 1987) spontaneously grew back. Is that normal?"[21] This $45 gallon of "Tuscan Whole Milk" is the object of over a thousand reviews and is now part of the pantheon of parodied Amazon products. Its popularity is evinced by the fifty-six customer images for the milk. Customers typically submit images to Amazon so that others can better appreciate the scale or color of an unboxed item; in this case, people got creative. Among dozens of images for the milk, one combines it with the most famous of all Amazon products: a t-shirt. In the image, a carton of milk is superimposed on a full moon above three howling wolves.

The story of the "Three Wolf Moon" t-shirt has been covered by *Business Week*, the BBC, and the *New York Times*. In 2008, law student Brian Govern was searching for school books when Amazon recommended a t-shirt with an image of three wolves howling at the moon. Without purchasing the item, he wrote a satirical review that concluded with a list of pros and cons: "Pros: Fits my girthy frame, has wolves on it, attracts women. Cons: Only 3 wolves (could probably use a few more on the 'guns'), cannot see wolves when sitting with arms crossed, wolves would have been better if they glowed in the dark."[22] Both the review and the product went viral: the shirt spent almost two hundred days on Amazon's Top 100 list and has been widely referenced in popular culture.

The review parody has now become an established genre of online comment, and most of the products that become topics of fun are absurd in some way. The "'Guardian Angel' Acupuncture Device" is absurd in its transparent effort to disguise its function as a sex toy. A banana slicer is absurd for the presumption that it is needed. Sometimes reviews reflect an absurdity of the larger social context, such as the unnecessary gendering of a ballpoint pen as "Elegant design—just for her." Sometimes the presumed context makes a product absurd. Because Amazon is seen as selling to ordinary consumers, its offer of a laparoscopic gastric bypass kit was parodied by many who recounted stories of at-home surgery. One widely discussed product evoked an absurdity of the modern condition: airport security checks. Although Playmobil hoped to lessen children's anxiety about security checks by allowing them to experience the process through play, A *New York Times* headline stated that "Playmobil's checkpoint strikes some as too real." Parody reviews complained that it was not real enough: the toy's luggage could not be opened and inspected, and its figures wore shoes that could not be removed.[23]

For most of its history, Amazon ignored funny reviews, neither condoning nor removing them, but in the summer of 2013, it finally recognized them on a page that stated "Helpful product reviews written by Amazon customers are the heart of Amazon.com, and we treasure the customers who work hard to write them. But occasionally customer creativity goes off the charts in the best possible way."[24] These funny reviews have even become the data on which researchers train their computers to detect irony.[25] In any case, beyond absurdity, sometimes the most confounding comments are those that arise from human foibles.

Oops: "This Is What Life Is Like in the World of Social Media"

In October 2011, Rainn Wilson made a mistake. The actor known for playing Dwight Schrute, a paper salesman and beet farmer on NBC's *The Office*, tweeted to his assistant: "Joanne—tell @DelTaco I will accept $12,000 to plug their sh***y food. Thanks, Rainn." A number of reports concluded that this faux pas was a failed direct message.[26] The mistake of publicly tweeting a private message is easy to make, one need only fail to prefix the message with the letter *d* (direct). The consequences of such a mistake can lead to more than embarrassment. A few months before Wilson's gaffe, Anthony Weiner was forced to resign from Congress for tweeting an inappropriate photo of himself to a young woman. The public tweet and close-up of his bulging underpants were quickly deleted, and the next day his spokesperson stated that Weiner's accounts had "obviously been hacked" and the story was a distraction from "important work representing his constituents."[27] Weiner said that "People get hacked all the time and it happened to me, I don't think it is the end of the world. This is what life is like in the world of social media and I will still be using Twitter as I think it helps me do my job." Eventually, he conceded that the photo could be of his crotch: "It could be. Or could have been a photo that was taken out of context or was changed and manipulated in some way."[28] Finally, he admitted to inappropriate online relationships, resigned from Congress, and apologized to all involved, including his wife. Two years later, while campaigning for the New York mayor's office, he had to apologize again—with his wife standing by his side—for having continued his sexting. In addition to being pathetic, this story touches on the bemusing issues of mistakes, context, and excuses.

In Twitter's early days, its creators were not sure about the model of openness and communication they should adopt. Was tweeting like emailing (a private exchange) or blogging (a public broadcast)? Journalist

Steven Levy spoke with some of Twitters' original designers, who recalled their initial uncertainty and their initial decision to keep users' profiles private by default.[29] As people became increasingly comfortable with public profiles and open exchanges, Twitter followed suit: users' tweets, followers, and followed were made accessible to others. The concession for private exchanges was that a person could directly message a follower by beginning a message with *d*—easy enough to forget.

Even beyond slips of the finger, people openly post appalling comments. These are not only shocking for their bigotry but for their candor. For instance, news of Obama's 2012 reelection was followed by some racist tweets. The pop-feminist blog *Jezebel* found that many of the tweets originated from teenagers whose "accounts feature their real names and advertise their participation in the sports programs at their respective high schools." Jezebel published information about some of the tweeters and "contacted their school's administrators with the hope that, if their educators were made aware of their students' ignorance, perhaps they could teach them about racial sensitivity."[30] This revelation of the teenagers' identities was controversial and raised an issue that people find confusing: context. *Jezebel* did not reveal anything about the tweeters that was not already available online, but the question was whether including these comments in a critical exposé somehow took them out of context.

As noted in earlier chapters, online comment is typically reactive, asynchronous, and short. Because comments are reacting to something, they are inherently contextual. Yet comment's asynchronicity and shortness often confuse readers as to what that context is. Online comment is often portable, as well, and transcends space and place as it is forwarded and retweeted. All of this often obscures the author's intention—an ephemeral human feature in the best of cases. Paul Chambers was annoyed when snow closed his local airport in England. He tweeted, "Crap! Robin Hood airport is closed. You've got a week and a bit to get your shit together, otherwise I'm blowing the airport sky high!!" Airport staff as well as two courts found the message to be sufficiently menacing to merit his arrest and conviction, but the decision was overturned by a high court panel that wrote that such a comment was hardly menacing when most would likely "brush it aside as a silly joke, or a joke in bad taste, or empty bombastic or ridiculous banter."[31]

In a related case in the United States, reason has yet to rule. Eigh-teen-year-old gamer Justin Carter had been arguing with friends on the League of Legends Facebook page, and when someone wrote that he was "insane" and "messed up in the head," he sarcastically agreed. His father, Jack Carter, claimed that his son responded: "Oh yeah, I'm real messed up in the head, I'm going to go shoot up a school full of kids and eat their still, beating hearts," followed by "LOL" (laugh out loud) and "JK" (just kidding).[32] Someone in Canada saw the message, found Justin's address, and reported him to the authorities. His father said, "These people are serious. They really want my son to go away to jail for a sarcastic com-ment that he made." Speaking to the temporal aspect of online comment, Jack stated that his son had been unaware of the Sandy Hook Elemen-tary School shooting a few months earlier and "was the kind of kid who didn't read the newspaper. He didn't watch television. He wasn't aware of current events." He also touched on the issue of place: "These kids, they don't realize what they're doing. They don't understand the implica-tions. They don't understand public space."[33] (Justin spent five months in jail, where he claimed to have been beaten and sexually assaulted, and was eventually freed on a $500,000 bond thanks to an anonymous donor. His trial has yet to start.) As noted in an earlier chapter, scholars would describe this as a collapse of contextual integrity: the trash-talking context of teenage boys had been lost. Perhaps this is why, in 2014, the U.S. Secret Service, which is charged with protecting government officials from harm, asked researchers for tools that can detect sarcasm and "false positives" in threatening online comments.[34]

The timing of comment has also prompted controversy. On the morn-ing following a shooting in a Colorado movie theater (during the Thurs-day midnight opening of the *The Dark Knight Rises*), a National Rifle Association (NRA) affiliate tweeted, "Good morning shooters. Happy Friday! Weekend plans?" The NRA removed the tweet and responded that "A single individual, unaware of events in Colorado, tweeted a com-ment that is being completely taken out of context."[35] The claim that the quote was "taken out of context" is an interesting one. For what context does a short comment, broadcast to the world, actually have? Is it the context understood by the sender or that used by the recipient? Com-munication theorists, notably Stewart Hall in his 1973 essay "Encoding/ Decoding," have long argued that information is not simply transmitted and received.[36] Instead, messages contribute to a constructed meaning that depends on various interpretative frames. Although comment in the age of the Web is often hypertextual (beyond textual), it is also often *hypotextual* (undertextual). By this I mean that the while the relationship between a digital comment and its object is often explicit, this link is easily broken as the message circulates. Additionally, people are often promiscuous in applying different frames to the message's interpretation. (Scholars label the multiple meanings of a signifier as *multivalent* or *polysemous*.)

Around the same time as the NRA incident, comedian Louis C.K. was accused of defending the offensive antics of fellow comedian Daniel Tosh, who allegedly made a rape joke during a performance to which a female audience member objected. When Tosh responded that she probably had (or should) be raped herself, a weeklong debate about rape culture, the prerogatives of comedians, and the ethics of taste ensued. Several comedians tweeted their support of Tosh, and Louis C.K. tweeted: "@danieltosh your show makes me laugh every time I watch it. And you have pretty eyes."[37] Digital comments are hypertextual in that they often include context-setting links. For instance, email messages have "In-Reply-To" and "References" headers. Blog postings can have links and trackbacks. Retweets are bound to the original tweet. (Though the meaning of a retweet is not always clear: some people include "IRT" in their message to indicate an ironic retweet, which ironically enough, has other meanings as well.) In this case, C.K. addressed Tosh by his username but did not use the hashtags circulating at the time that designated the rape-joke context. For instance, in a follow-up, Tosh tweeted, "the point i was making before i was heckled is there are awful things in the world but you can still make jokes about them. #deadbabies."[38] This hashtag signifies the context of a discussion about the ethics and prerogatives of making jokes about awful things. (This tag is used to designate many other contexts as well, including abortion.) Those who objected to Tosh's behavior used the tag #ToshPointNo. However, C.K. told Jon Stewart on *The Daily Show* that he was unaware of the ongoing controversy:

I was in Vermont and I was watching TV in a hotel room and Daniel Tosh's show comes on. It's making me laugh; it's a funny show. So, I wasn't reading the Internet at the time because that's how I go on vacation. I really hate the Internet, so I just stopped reading it. But I'm watching TV and Tosh is making me laugh, so I wrote a tweet to say "Your show makes me laugh." And then I put it down, and two days later I come home and I read these bloggers and *Hollywood Reporter*:

"Louie CK Defends Daniel Tosh Amid Rape Joke Controversy." I had no idea he got in trouble for making some jokes about rape![39]

Tosh later claimed that he had been misquoted and tweeted: "All the out of context misquotes aside, I'd like to sincerely apologize."[40] But as is often the case with tweeted apologies, what exactly is being apologized for is not clear.

Biases: "F.ck me! This Is a Babe?!!"

Another consequence of the shortness and possible immediacy of online comment is that we reveal our prejudices in ways that we would not otherwise. Quick responses may act like an *implicit association test*, a research tool that reveals cognitive bias. People who take the test are asked to quickly associate terms with a category. The revealing part of the test comes when participants are asked to associate a term with a combined category. For instance, to associate "successful" with a "black/rich" and "white/poor" and later with "white/rich" and "black/poor." Over many iterations, over different terms and categories, the split-second differences between associations reveal implicit biases despite any conscious efforts. For instance, in the United States, many people (even people of color) are quicker to pair "successful" with "white/rich" than "black/rich."[41] Often tested stereotypes include age, race, and gender, but even marketers use it to discern preferences among soft drinks.

I sometimes think that Twitter serves as an implicit association test. When the creator of the popular Facebook page "I Fucking Love Science" posted a link to her Twitter page (that included a profile picture), some of the reactions revealed implicit assumptions. Elise Andrew had never hidden her gender, nonetheless some of her four million fans betrayed their bias: "F.ck me! This is a babe?!!"[42] Similarly, in response to the 2012 film adaptation of the book *The Hunger Games*, readers complained about the casting of the actors: "I was pumped about the Hunger Games. Until I learned that a black girl was playing Rue." This is one of the least bigoted tweets among dozens of surprised reactions, even though Rue is described in the book as a twelve-year-old girl with "dark brown skin and eyes."[43]

Sometimes the bias isn't even that implicit, though it is still surprising that people think their sentiments are appropriate to a public audience. In 2013, evolutionary psychologist Geoffrey Miller tweeted, "Dear obese

PhD applicants: if you didn't have the willpower to stop eating carbs, you won't have the willpower to do a dissertation. #truth." When he was challenged for this he responded that finishing a dissertation is "about willpower/conscientiousness, not just smarts." However, his ill-conceived (and not very smart) tweet prompted much criticism. Or, rather, many of our thoughts are ill-conceived and online comment provides few barriers to their expression. As public attention to the incident grew, he explained that his comment was part of a study measuring the reaction to provocative tweets. Yet this rang false: university researchers must conduct such "human subject" research under strict ethical guidelines and institutional review. As his comments increased in notoriety, he deleted the offending tweet, issued an apology, and made his account private. He wrote "My sincere apologies to all for that idiotic, impulsive and badly judged tweet. It does not reflect my true views, values or standards." One online response to Miller's tweet was the creation of a blog with almost a hundred self-submitted photos: *Fuck yeah! Fat PhDs: Being Fatlicious in Academia.* An institutional response followed a few months later, Miller was censured by his university, which required his continuing supervision and recusal from graduate admissions.[44] Will power and impulsivity are not as straightforward as we'd like to think, especially under the influence of Twitter.

Excuses: "I Was Hacked!"

The idea that things can be taken out of context is widely enough recognized that Anthony Weiner used it as an excuse for his crotch shot. He also used the excuse of being hacked. (Even after Weiner admitted to his compulsive sexting, some conspiratorial partisans persisted in the belief that he was hacked and argued that he was being framed and his confession was coerced.) This lament, "I've been hacked," is an often repeated refrain. Bloggers at *Deadspin* compiled a list of over a dozen such claims from sports figures alone: "Rich Eisen was not 'so horny.' He was hacked.... Rasual Butler did not Tweet his penis. He was hacked.... Santonio Holmes did not tell a fan to kill themselves. He was hacked.... Ron Artest did not have it up to here with Phil Jackson. He was hacked.... Tito Ortiz did not post a photo of himself 'wearing nothing but a smile.' He was hacked."[45]

Another famous deployment of the "hacked" claim was from the owners of Amy's Baking Company (ABC). In May 2013, the ABC restaurant was featured in an episode of Gordon Ramsay's *Kitchen Nightmares*. Despite Ramsay's often confrontational manner, the show's structure typically resolves early battles with Ramsay into a redemptive story of a successful restaurant relaunch. Yet ABC has the distinction of being the sole restaurant from which Ramsay walked away.

The proprietors, Amy and Samy Bouzaglo, told Ramsay that they had been struggling with negative online reviews and stated they were proud to stand up to "online bullies and haters" including "Yelpers" (the nickname for contributors to Yelp). Their defensiveness is not unique; everyone can be exploitative in the ratings game and customers are no different, as seen on blogs such as *Fuck You Yelper* and *Yelpers Suck*.[46] Amy and Samy hoped that the show would help them to showcase their establishment. Although Ramsay appreciated the dessert he offered critiques of the dinner and service: the "freshly made" ravioli was purchased frozen, and service was slow because Samy insisted on entering all orders himself. The Bouzaglos did not react well to the criticism and claimed that during one of their blowups the dining room had been full of haters. In a subsequent interview, Amy recalled that "I grabbed the producers in the middle of filming, I said, 'They are Yelpers. I know two of them for sure.' ... My husband and I started to feel that we were surrounded by Yelpers, completely set up.... the people that my husband is screaming those obscenities at, they are not our customers. They are Yelpers.... All of those people, they went there with harmful, malicious intent."[47]

When the show aired, it became an online viral phenomenon and a Reddit thread on the topic was especially popular as were fake reviews and parodies. In response, the social media accounts associated with ABC began issuing bizarre declamations. In reference to the frozen ravioli, one comment stated: "I AM NOT STUPID ALL OF YOU ARE. YOU JUST DO NOT KNOW GOOD FOOD. IT IS NOT UNCOMMON TO RESELL THINGS WALMART DOES NOT MAKE THEIR ELECTRONICS OR TOYS SO LAY OFF!!!!" Indicating a lack of online savvy, another comment warned the "Yelpers and Reddits" that "THIS IS MY FACE-BOOK, AND I AM NOT ALLOWING YOU TO USE MY COMPANY ON YOUR HATE FILLED PAGE." The comments were characterized by capitalized and broken English, bravado, obscenities, and mentions of

God: "WE ARE NOT FREAKING OUT. WE DO NOT CARE ABOUT A 'WITCH HUNT' I AM NOT A WITCH. I AM GODS CHILD. PISS OFF ALL OF YOU. FUCK REDDITS, FUCK YELP AND FUCK ALL OF YOU. BRING IT. WE WILL FIGHT BACK." This added fuel to the fire. Eventually, a measure of sanity was established with the following post: "Obviously our Facebook, YELP, Twitter, and Website have been hacked. We are working with the local authorities as well as the FBI computer crimes unit to ensure this does not happen again. We did not post those horrible things. Thank you Amy&Samy."[48] (Many questioned the simultaneous compromise of all their accounts and their invocation of the FBI.) This strange story did not end with this comment: the restaurant owners tried to capitalize on their infamy by selling t-shirts ("Here's your pizza, go F**K yourself") and there was talk of the restaurant having its own reality show.

Of course, some accounts are hacked, which can have an impact beyond personal embarrassment. When Associated Press's account was commandeered to send a dozen words to its almost two million followers, the S&P 500 briefly lost over $100 billion in value: "Breaking: Two Explosions in the White House and Barack Obama is injured."[49] In most cases, though, only the tweeters themselves and perhaps Twitter know which of these excuses are genuine. Claims of hacking have become another weird aspect of online culture. In fact, a few months after ABC's meltdown, Chipotle Mexican Grill "hacked" its own Twitter account as a publicity stunt.[50] WTF.

Whisperers: "Weird SENSATION FEELS GOOD"

Comment has been with us since the advent of writing. Hating, liking, and manipulating have been with us even longer. Robin Dunbar argues that these behaviors are associated with our origins as large-brained social animals. What is unique in the age of the Web is that people can comment from the living room, office, and street via clicks, text, images, audio, and video. This enables a degree of ubiquity and scale never seen before. Sometimes these features lead to novel manifestations of comment, like the unboxing and haul videos discussed in an earlier chapter. Comment's reactivity and quickness also lead to accidents and revelations of stupidity.

People also can learn something new about themselves at the "big data" scale of thousands and millions of comments. "Internet geographer" Monica Stephens has mapped the racist, homophobic, and ablist slurs that appeared in over a hundred thousand geocoded tweets. She was surprised to find that although the word *nigger* was used in the southern states of Georgia and Alabama, its use also was concentrated in small towns in the Midwest and Rust Belt.[51] Another mapping project sought to identify possible food poisoning incidents, most of which go unreported. Millions of New York City geocoded tweets were scanned for allusions to sickness (such as "threw up") and associated with nearby restaurants. It identified likely outbreaks of food poisonings and was able to estimate nearby restaurants' food safety grades.[52]

The ubiquity and scale of today's comment also lead to a greater appreciation of what is referred to as the *long tail*, the many niches that previously were underserved. Amazon sells blockbuster books but also thousands of lesser selling titles. Gay teens who feel isolated in their small towns can reach out to others. Those suffering with rare diseases can communicate about and advocate for their concerns. The troll attacks on the Epilepsy Foundation's forums were particularly insidious because they attacked a means of connecting otherwise isolated people.

Sometimes comment at the scale of the Web even facilitates the emergence of something new, such as what is now called the autonomous sensory meridian response (ASMR). The site Is It Normal? is another Q&A Website, but instead of commenters asking if they are hot or ugly, they typically describe odd, curious, and often embarrassing phenomenon and ask if others have experienced the same. In April 2009, someone wrote about an odd response to an unboxing video:

Ok basically i was watching the video on youtube and it was about unwrapping this package my friend got (no im not saying what the package was) and well I'm not sure how to describe it really its hard to explain but at the back of my head I sort of get this internal massage but its a really nice feeling and it sort of make me go numb so if im typing I lose feeling in my fingers its weird but its REALLY soothing and calming I sorta dont want the moment to end HELP![53]

There are now almost three hundred comments, many of which describe triggers (such as watching people draw, fold paper, or talk quietly) and a pleasant tingling sensation. Many comments are epiphanous: "Yes yes

yes!!! OMG I Can't believe I found other people with this!! *dies* I never even posted about it because I can't fucking describe it!" In early 2010, the phenomenon became a popular topic on the site Steady Health. The thread "Weird SENSATION FEELS GOOD" appeared in the "Nervous System Disorders and Diseases" forum, indicating that while pleasant the seeming solitary condition could also be a source of anxiety. Yet, apparently, others had been talking about this phenomenon for a few years under the monikers of AIHO (Attention Induced Head Orgasm) and AIE (Attention Induced Euphoria). The Website AIHO.org read: "Have you ever had that weird tingling sensation on your scalp when watching someone do something, or hearing them speak? That sensation which feels really good.... You aren't the only one!"[54]

By the summer of 2010, the community of "sensationalists" had reached a critical mass and launched projects to research autonomous sensory meridian response (ASMR), including its purposeful inducement. The blog *The Unnamed Feeling* became a popular site for chronicling events within the community, and an ASMR playlist was posted on You-Tube. In 2012, ASMR was widely recognized when it appeared on the reference sites Wikipedia and Know Your Meme.[55]

In 2013, writer Andrea Seigel appeared on Public Radio International's *This American Life* to describe her experiences with the sensation and her discovery that it was something that others had experienced. She confessed that she spent hours a day on YouTube and was especially fond of jewelry and cosmetic haul videos. She loved "someone speaking in lightly accented English" and the "tapping of a brush on a Mac pigment bottle, or the clicking up of an eyeliner pencil." One night, at a loss for new videos, she thought that because many speakers with accents had difficulty saying the word *jewelry*, perhaps she should search for misspellings of the word. The top result was the video "ASMR, Old Jewellry Collection, Show and Tell Whisperer" posted by the user TheWaterwhispers. She was stunned to find that her weird and personal predilection was shared by others: "In an instant, I went from believing I was miswired to suddenly feeling like I was part of a special group of people with amazing sensitivities."[56]

In addition to how comment informs and improves, alienates, manipulates, and shapes, these short and asynchronous messages can bemuse:

they can be slap-dash, confusing, amusing, revealing and weird. However, from this confusion and weirdness, we can learn about the advantages of moving first, the challenges of communication, the science of rating systems, and the importance of context at the bottom of the Web. In particular, because comment is reactive, it is inherently contextual; it also is hypotextual, shedding context with ease, which prompts the retort of "WTF?!?" in response.

8

Conclusion: "Commenterrible"?

The Internet took commentary and made it commenterrible.
—@AvoidComments (Shane Liesegang), Twitter

Amanda Brennan describes herself as "a complete nerd" who is passionate about absorbing information and sharing it with others. She even has managed to make a career of it and served as a "meme librarian" at the Website Know Your Meme and, more recently, a "Tumblarian" at the popular microblogging site. She also has an interesting piece of jewelry: a "never read the comments" necklace. However, is comment really so "commenterrible" that we should never read it? For those with maximizer tendencies or anyone else who reads an online review, comments that inform have obvious utility; it can be thrilling to have so much information available. Yet the bonanza of information and choice can also become a glut that paralyzes people who worry about purchasing a $5 item without first checking its online reviews. Online shoppers call the moment that their anxious ruminations finally break as "pulling the trigger" for good reason: whatever the hesitations or consequences, the deal is done. Even putting aside the utility of reviews, those who stylishly repudiate comments are still likely to read them. After all, Brennan's job is to read the comments. So the maxim is not really a prohibition but a comment on the sometimes sorry state of comment in an otherwise compelling medium. The usefulness of the Web is demonstrated by where Brennan first saw the "never read the comments" necklace (on Instagram) and where she bought it (on Etsy). She says that for her, the necklace means that the Web can still be a place of thoughtfulness and creativity:

"Knowing that other people out there get that is a reminder that not everyone wants to use the internet in its darkest ways."[1]

One way of thinking about reading online comment is that it is like watching the television game show *Family Feud*, where two families compete to answer questions quickly. The goal of the game is not to give a correct answer but to provide an answer that corresponds with what others have said, what the "survey says." A contestant who is asked to "name a U.S. president whose face appears on money" might answer "Benjamin Franklin," and although Franklin never was president, enough people think so as to make it a winning response. Similarly, insight and wisdom might not always be found at the bottom half of the Web, but it does have a sample of what some people are thinking, right or wrong, offensive or trite. As much as we may dislike or wish to distance ourselves from comment, I often think of Walt Kelly's decades old *Pogo* comic strip: we have met the commenter, and "he is us."

Just as in the experiments in which participants who are exposed to a mirror feel a little worse about themselves, we sometimes prefer not to look into the online reflecting glass of humanity. Some people disable or restrict comment on their sites. The popular blog *Boing Boing* has long wrestled with comments. It disabled discussion altogether in 2003 and reintroduced it in 2007 with the hiring of a community manager. Its new philosophy was captured in a 2008 essay by *Boing Boing* contributor Xeni Jardin, who wrote that "Online Communities Rot without Daily Tending by Human Hands." And those human hands had a new tool: the "disemvoweller" removed all the vowels from the screeds of those misbehaving. Jardin wrote that "The dialogue stays, but the misanthrope looks ridiculous, and the emotional sting is neutralized."[2] Yet this technique was a passing fad: it did not last, and unruly comment did. Although such pillory can be appealing, it can backfire. In a study of millions of comments across four sites, including CNN, researchers found that commenters who were down-voted subsequently posted more and lower-quality comments. Those who were down-voted also were more likely to reciprocate in kind, "percolating these undesired effects through the community."[3]

Boing Boing's comments continued to be problematic, and in early 2013, Rob Beschizza, the blog's managing editor, tweeted that he might disable comments permanently. Instead, later in the summer they tried again and adopted a platform that "acts as a neat hybrid of forums and comments. It's designed to offer the most useful features of a modern discussion platform, yet remain simple to read and easy to use for everyday readers."[4] With the new Discourse system, comments can be made off the main page, and a moderator can promote select comments to the story's page. Many other sites are adopting this model by featuring curated comments below a story. At sites like the *New York Times*, comments are accepted on select articles for a limited duration. Yet since the switch to the new comment policy at *Boing Boing*, each post only provokes a few comments and a larger discussion rarely has more than a dozen. Moderators rarely, if ever, feature any "promoted" comments below a story. As one *Boing Boing* commenter noted: "Rob's been quite open in the past about not liking the commenters. I think this is just a bridge-step to remove them from the site, disassociate if you will. Having used Discourse for a little while now it should be incredibly effective at silencing conversation."[5]

As hard a nut as it is to crack, people are still trying to solve the comment problem. Lawrence Lessig, who in 2009 temporarily abandoned his blog because of link spammers, recently enthused about the Website Medium.com, where comments are like annotations, short and specific to a paragraph. He hoped that this would support the "beginning of a conversation with readers, much more seamlessly and effectively than the standard post+with+flamewar+in+comment+section pattern of blogs." This was "in theory, at least" because "some bug is blocking my ability to comment back." As evidence of his dedication, he wrote about his experience with Medium.com (blog #3) on Lessig.org (blog #1) via a reblog from his Tumblr (blog #2) and a parallel tweet.

More generally (and like Discourse), Gawker Media (including blogs such as *Gizmodo*, *Lifehacker*, and *Jezebel*) is experimenting with a hybrid system. With Kinja, users are given their own Tumblr-like blog that collects all of their comments, responses, and posts on Gawker-affiliated sites: "readers will then be able to use Kinja as a central hub for discussion on these stories, almost like their own chat room protected from the commenting maelstrom."[6] Similarly, those behind Discourse, which silenced much of the raucous conversation that once existed at *Boing Boing*, seek to quiet the maelstrom. Jeff Atwood, the cofounder of Discourse, has a surprising take on online discussion based on his earlier successes at Stack Exchange, a popular informational Q&A site:

At Stack Exchange, one of the tricky things we learned about Q&A is that if your goal is to have an excellent signal to noise ratio, you must suppress discussion. Stack Exchange only supports the absolute minimum amount of discussion necessary to produce great questions and great answers. That's why answers get constantly re-ordered by votes, that's why comments have limited formatting and length and only a few display, and so forth. Almost every design decision we made was informed by our desire to push discussion down, to inhibit it in every way we could. Spare us the long-winded diatribe, just answer the damn question already.[7]

While Atwood does not need to be quite so severe at Discourse, the notion that less is more persists. In a separate post entitled "Please Read the Comments," Atwood wrote that "if you are unwilling to moderate your online community, you don't deserve to have an online community. There's no end of websites recreating the glorious 'no stupid rules' libertarian paradise documented in the *Lord of the Flies* in their comment sections." He concluded that such a site ends up "exactly as you would expect it to."[8] Although Kathy Sierra was harassed in 2007 for recommending that bloggers should moderate their comments (harassment that began at the *Mean Kids* blog, an example of a "no stupid rules" blog), moderation is now a central tenant of successful blogs. In a talk at MIT's Center for Civic Media, Ta-Nehisi Coates (a senior editor of and blogger at *The Atlantic*) said that seeing the following below a story was inimical to creating a community: "5,000 comments. Join the conversation." Instead, limiting comment should be thought of as cultivating a garden: "Once you take out the rubbish and clear away the weeds, flowers begin to grow."[9]

Beyond the new hybrid systems and a philosophy of pruning the weeds, we could be even more imaginative. Judith Donath's recent book, *The Social Machine: Designs for Living Online*, details a decade's worth of research aimed at understanding the workings of online interaction. It includes visualizations of online conversations that show the health of a community and who is playing what role (for example, leaders, contributors, and cranks). This approach could help us better understand online discussion as something organic rather than as an experimental "libertarian paradise." Or imagine that in the place of a profile picture, a comment section instead showed a "data portrait" that represented a commenter's most salient or frequently used words.[10] (Such portraits or a display of the number and average rating of an Amazon reviewer would be useful.) Time will tell if such ideas could help, but many large sites are pursuing a different path.

Despite Clay Shirky's insight that "comment systems can be good, big, cheap—pick two," the larger sites are still trying to have it all. They hope to achieve this via identifiable users, single sign-ons, and the social graph. Although such efforts *might* improve online discussion, they certainly benefit their proprietors (and advertisers) with much richer information about users. And it is by no means certain that such efforts will yield Shirky's triple crown. In late 2013, YouTube launched a new initiative at "turning comments into conversations that matter to you." How would they accomplish this? They would integrate Google+ into YouTube. (I imagine they also hoped that this would make Google+ more competitive with Facebook.) On the commenters' side, Google linked YouTube comments to an identifiable Google+ profile. Also, because YouTube comments were now Google+ comments, they lacked some of the restrictions that YouTube had previously used to limit abuse: comments now could contain links and were no longer limited in size. On the reader's side, they would be more likely to see comments from people who were in their Google+ circles, from known "YouTube personalities," and from those comments receiving the most engagement. Google was borrowing Facebook's approach wherein Facebook decides how many friends see a post. Google seemed confident that identifiable comments from those who were embedded within a social network would improve YouTube's notorious comment culture.

Under the new system, however, comments with high engagement tended to be the most inflammatory, making it a target for trolls. The ability to include links yielded an influx of spammers, and the change annoyed many, especially YouTube users who resented having Google+ forced on them. Worse yet, Google's push to integrate users' accounts ended up "outing" some who previously had relied on pseudonymity for safety. In borrowing Facebook's approach to engagement, Google alienated many of its YouTube users. Over 200,000 people signed a petition in protest. Even YouTube's cofounder, Jawed Karim, asked "why the fuck do i need a google+ account to comment on a video?"[11] The hundreds of comments below Google's announcement were illustrative of the reaction. The top comment asked, "Who the fuck are YOU to decide who's comment I want to see? This condescending attitude is extremely annoying." Another noted, "that's nice your fixing up spam" but "I don't want youtube to be a social network! If I wanted to go on a social network, I would join facebook!"[12] In a fit of pique, YouTubers took advantage of the fact that there were no character limits for comments and posted massive renditions of ASCII-art; textual penises proliferated. Embarrassed, Google promised that it would fix the new system with "better recognition of bad links and impersonation attempts," "improved ASCII art detection," and "changing how long comments are displayed."[13] The worst abuses eventually abated, but in April 2014, the head of Google+ announced his departure, and it appeared that Google would be relocating many of its employees away from Google+ integration.

A more deeply felt disappointment of 2014 was that MetaFilter, the long-lived example of successful online discussion, announced that it was laying off three of its eight staff, including a long-time moderator. Its skillful human moderation, a one-time $5 membership fee, and strong community norms had continued to foster a successful comment culture. Yet the revenue it depended on from advertisements placed in its informational Q&A section "Ask MetaFilter" was declining. These pages had long been prominent in Google search results, but something changed in October 2012, and traffic abruptly dropped by 40 percent. Many suspected that Google's continuing battle against link spammers led to an algorithm change that resulted in MetaFilter's many hypertextual comments being seen as spam.[14]

I fear that the future of online comment will continue to move toward large commercial platforms in which people have little privacy and see mainly the posts of the likeminded, the popular, and those who pay to reach us—a neutered filter bubble that serves advertisers rather than users.

I began this book with stories of those who were fleeing filtered sludge— that is, online discussion that failed to scale in the face of scammers and haters. I argued that this implies two things. First, commenting systems will forever be attempting to fortify themselves against abuse. Second, new commenting platforms will continue to appear as people will move in search of intimate serendipity, a place where they can express an authentic sense of self without fear of attack, manipulation, or unusual exposure while remaining open to things that will surprise and delight them. A third option is that even if the comment system is enabled, some people will simply pull back. *Boing Boing* contributor Xeni Jardin, who "Instagram[ed] my mammogram," later wrote that although she found it comforting to share her distress she now appreciated a need for balance: "that just as there is value in connecting, there can also be value in disconnecting and just dealing with what's going on inside our bodies and inside our minds."[15] Mark Frauenfelder, who began the *Boing Boing* blog with his wife, Carla Sinclair, almost twenty years ago, told me that "my view of online community has changed over time":

I'm no longer interested in responding to comments to my posts. The subtleties of face-to-face communication are lost. Most people are not polite online, including me. I get upset with myself when I become snarky. The purpose of my posts on *Boing Boing* isn't to create a conversation online. It is to point to things that interest me. If people are compelled to make remarks about my posts, they are free to do so in our comments section, their own blog, or on a 3 × 5 index card pinned to a laundromat corkboard. I stopped participating in the comments section of *Boing Boing* a couple of months ago. I feel better because of it.[16]

In 2012, Dave Winer, the person who often is credited with first deploying blog comments, disabled them, and it seems that many long-time bloggers have come to share Winer's sentiment that comments are sometimes valuable but, on balance, "they're not worth the trouble."[17] Even so, a new generation of people and platforms is always on the horizon. When Twitter first launched, it seemed like a high-signal and low-noise sort of place that could afford its members intimate serendipity. In the face of success, however, this achievement rarely lasts. As Trent Reznor of the band Nine Inch Nails learned, intimate serendipity can soon turn to wading through "sludge," and he now restricts himself to a few tweets a month about his work. Other celebrities swing between manic engagement and quitting. Miley Cyrus's first quit (among many) was when she was sixteen. Reznor's disillusionment from 2009 persists:

I was excited about Twitter when we went out on our own because it felt like the most direct way to penetrate people's attention. I also got a charge out of people realizing that I wasn't a recluse sleeping in a coffin. But in hindsight, my experimenting with Twitter was a mistake. Oversharing feels vulgar to me now. I know we've been fooled into thinking it's okay to show dick pics and that the Kardashi-ans' behavior is normal, but it's not. I've tuned out in the last couple years. Every-body's got a fucking opinion. It takes courage to put something out creatively into the world, and then to see it get trampled on by cunts? It's destructive.[18]

Reznor's words are crude, but it does take courage to put one's self and creative works out there for judgment, which leads to the question of how to best exchange feedback (comment that is intended to help others improve).

Many people find it worthwhile to receive feedback on their creative endeavors, even though they might receive harsh criticism. Giving criti-cal feedback also can be difficult, as was the case for the Toastmasters's participant who felt nervous about evaluating a brilliant speaker. As has been shown in the cases of authors and their readers, even skillful feed-back sometimes prompts angry responses. Purposeful parody of amateur fiction by way of *sporking* (reviewing work that is so bad that reviewers want to spoon their eyes out) is almost guaranteed to hurt feelings if it gets back to the author. (The author of what has been called "the worst fanfiction ever" actually might be pleased. *My Immortal* was widely sus-pected to have been a purposefully awful story of a teenage vampire who attends Harry Potter's alma mater, Hogwarts.) The scope and scale of comment also have changed online. In the past, feedback's scope was relatively local. In a public speaking group, candid feedback from mem-bers is shared with the speaker or at least remains within the confines of the room. Online, comment is *hypotextual*, and unsolicited comment can easily find its way to the subject—and to everyone else.

Additionally, online communication lends itself to what I call *drama genres* of comment. In a lecture about television in the 1970s, media theo-rist Raymond Williams observed that "We never have as a society acted so much or watched so many others acting." To our ancestors, drama was periodic, as experienced in the celebration of a religious festival or the occasional "taking in a show." According to Williams, in the age of tele-vision, drama "is built-in to the rhythms of everyday life. What we have now is drama as habitual experience: more in a week, in many cases, than most human beings would previously have seen in a lifetime."[19] Online, people can see more drama in a week than Williams could see in a lifetime of watching television. And I use the word *drama* in two senses—scholars' notion of performance and teenagers' sense of histrionics. In the age of comment, people are always performing front of stage, and much of it is sensational.

Specific *drama genres* of comment include personal Q&A sites and online lists. Who would've imagined that being able to make a list of books at Goodreads would be so contentious? Since the *Whole Earth Catalog* of the 1960s, people (that is, likers) have shared their lists and recommendations for the benefit of others: "Here are some things that I've enjoyed or that have improved my life. Perhaps they can do so for you too." The blog *Cool Tools*, a successor to the *Whole Earth Catalog*, recently featured an "invisible glove" lotion that protects the skin. Painters and mechanics use it to easily wash off paint and oil, but the commenter noted that it also was useful for protecting skin from poison oak while hiking.[20] Handy! People similarly like to share their favorite authors and books, but at Goodreads, they also circulated "do not read" lists of out-of-favor authors. Some authors felt that the lists were a form of bullying and compiled their own lists of "Goodreads bullies," which was also rather bullying. These *bully battles,* in which factions compile reciprocating lists of bullies, also have appeared on Twitter with the sharing of automated block lists. Of course, "kill files" were used on Internet discussion forums long before the Web. Back then, to publicly state that you were adding someone to your kill-file or bozo-filter was a public insult, but one which tended to tamp down on the flames because they hid others' postings, including their angry responses. Today, the use of lists seems much more factionary and inflaming.

Similarly, the Q&A genre is ripe for drama and abuse, especially when questions can be posted anonymously. Casey Newton, senior reporter for *The Verge*, addressed this topic in an article with the sadly clever title "Killer App: Why Do Anonymous Q&A Networks Keep Leading to Suicides?" The article recounted Formspring's initial popularity, its struggles to deal with abusive anonymous content, and its eventual eclipse by Ask.fm, which also has had its share of teenage suicides. When Formspring shut down in March 2013, its lead designer, Cap Watkins, reflected on these struggles and concluded that although anonymous Q&A was responsible for the site's initial success they had had "protected anonymous content to a fault":

On the one hand, anonymity was a really popular feature (duh). On the other hand, we saw a lot of bad and abusive content come through that channel (double duh). A fact that we wound up being pretty infamous for. But man was it hard to let go of anonymity as a core feature. We tried workaround after workaround. We prompted for sign-up after asking an anonymous question. We started pushing privacy settings for users into our on-boarding (which they never changed, of course). We started setting up elaborate filters to catch bad or abusive questions and put them behind a "Flagged Questions" link in users' inboxes.[21]

Formspring tried to hold onto the Q&A feature because it was a source of the site's popularity, but as it attempted to curb abuse, teens moved to its Latvia-based competitor, Ask.fm. As a former Formspring executive concluded in the "Killer App" article, "When you took out the nasty, salacious, anonymous part of Formspring, it became a lot less interesting to people."[22]

There is a lot of nasty comment out there, and it tends to be surprising in two ways. First is the extent of the awfulness, in severity and scale. In July 2013, the Bank of England announced that Jane Austen's portrait would be replacing that of Charles Darwin on the back of the £10 note. Caroline Criado-Perez, who had campaigned for increased female representation on the currency, received much of the credit for the change, and she also received a torrent of hate via Twitter. She and other advocates received what is now (unfortunately) to be expected: graphic death and rape threats. It was reported that the women received as many as fifty threats per hour. (She also received support, and a petition with over a hundred thousand signatures prompted Twitter to apologize for the abuse and to deploy a new abuse report system.) Comment can be used to express hate or support, but I suspect the deluge of hate leaves a much stronger impression than even the kindest expressions of encouragement. The police followed up with arrests, but Criado-Perez found their procedures (and possible loss of evidence) to be frustrating and traumatic: "They were now asking me to go through all the threats I'd received— and relive all the psychological trauma involved—to look for three specific usernames, to see what evidence I had of their abuse."[23] Despite the actions of Twitter and the police, a new round of threats led her to delete her Twitter account. This case also reflects the inescapable gender dimension of online alienation and hate, which has been popularly recognized in a number of aphorisms, including one I coined myself. In response to the attacks on Anita Sarkeesian for her *Tropes vs. Women in Video*

Games project, I noted that "Online discussion of sexism or misogyny quickly results in disproportionate displays of sexism and misogyny."[24]

The second surprising thing about embarrassing or nasty comments is that they often are made in the open and beyond the cover of anonymity. The shortness and ease of online comment often prompt people to make stupid mistakes and reveal implicit biases. Anthony Weiner surely regrets his mistake of failing to prepend his infamous tweet with the letter *d* to make the photo of his privates private. But why take such risks in the first place? And those who complained that *The Hunger Games* was ruined because a female character (who is described in the book as having dark skin) was played by a black actor revealed their bias. They might have thought that their opinions were being expressed only to their peers, forgetting that Twitter is public. To return to Goffman's metaphor of the stage, we are always in the presence of and performing for others in the age of ubiquitous comment. Keeping this in mind is a challenge for some people, who when they fail, can only claim "I was hacked!"

The legal system and lawyers are slowly recognizing the substance and import of online comment, but this is happening in fits and starts. In one case, former New Orleans police officers were accused of shooting unarmed civilians as they fled across a bridge to escape the devastation of Hurricane Katrina. As jury selection for the trial was beginning, an online commenter expressed his outrage toward the police on a newspaper's Website: "NONE of these guys should have ever been given a badge." This commenter was in fact a top federal prosecutor and was joined in the sockpuppetry by two other Department of Justice officials. When this came to light, the judge was forced to dismiss the convictions of the former cops, pointing out the sad irony of the "accusers becoming the accused" in such "grotesque prosecutorial misconduct."[25] The prosecutors recognized the ability of online comment to influence public opinion but did so in an inappropriate, if not illegal, way.

Courts themselves have been uncertain about the meaning and protections that should be afforded to comment. During a 2009 sheriff election, employees of the Hampton, Virginia, sheriff's office *liked* the Facebook page of their boss's opponent and were subsequently fired, which led to a court case in which the judge ruled that "merely 'liking' a Facebook page is insufficient speech to merit constitutional protection." The employees would not be reinstated. A number of media savvy lawyers questioned the ruling by asking why wearing a black armband is substantive and protected speech but Facebook liking is not. The U.S. Court of Appeals for the Fourth Circuit eventually reversed the ruling, writing that "On the most basic level, clicking on the 'like' button literally causes to be published the statement that the User 'likes' something, which is itself a substantive statement." Additionally, a *like* is a "symbolic expression" that "is the Internet equivalent of displaying a political sign in one's front yard, which the Supreme Court has held is substantive speech."[26]

In the realm of online reviews, the law is starting to take notice as well. In September 2013, the New York State attorney general's office announced that it had compelled nineteen companies to stop writing fake online reviews and to pay more than $350,000 in penalties. Posing as a yogurt shop with some negative online reviews, the office contacted search engine optimization firms that offered to post fake reviews through manufactured and obscured identities. "Operation Clean Turf" found that many companies were willing to post fake reviews on Yelp, Google Local, and Citysearch by way of "freelance writers from as far away as the Philippines, Bangladesh and Eastern Europe for $1 to $10 per review."[27]

Perhaps the most important aspect of online comment is much more modest than eye-catching headlines and closely watched court cases: it is comment's seemingly innocuous ubiquity. We live in a world in which everything can be commented on, and these data can be easily tallied. This is part of what I call *quantification*. As sociologist George Ritzer notes, people who live in contemporary "rational" societies are driven toward quantifiable measures, in part, because they rely on computers, which also make it easy to make difficult decisions when assessing subjective and qualitative phenomena. Additionally, in a winner-take-all society, it is no longer sufficient to be good enough and to be appreciated as such.[28] Instead, we are presented with a proliferation of identities to choose from, to perform, and to be judged on relative to world-class successes and beauties. Feedback (identifying what works well and what can be improved) is replaced by rankings of standing relative to others.

This preoccupation with identity, attractiveness, and relative standing was present at the birth of social media. "HOT or NOT" brought the Web to popular attention and inspired the developers of YouTube, which now hosts "Am I ugly?" videos. Facemash, Facebook's predecessor, was created by Zuckerberg as a way for Harvard students to judge the attractiveness of fellow students. In Katherine Losse's memoir of her time as an early employee of Facebook, she described her low status as a female customer support representative relative to the "boy king" engineers who had a fondness for quantification. She wrote that the launch of the 2007 Facebook Platform app was accompanied by an internal example app named "JudgeBook": "never judge a {face}book by her cover." Her colleagues even acquired the domain names JudgeBook.com and PrettyOr-Witty.com, which echoed a comment she had heard Zuckerberg make about having to choose between women who were pretty or witty. She felt that "in either case, you would definitely be judged, scored, and rated":

It was at moments like these that I realized it was the great and twisted genius of Facebook for anyone who is interested in rating things constantly, as Mark and the engineers who made these type of applications seem to love doing. Facebook made it possible for men to have endless photographs of women available for judging, and women simply by being on Facebook became fodder for the judging, like so many swimsuit models at a Miss America pageant. Because, with Judge-Book, like all Facebook platform applications, women did not have to consent to have their photographs used by the application. The application would alight upon your data and feed it into its database whether you wanted to be judged or not.[29]

Inescapable judgment is a consequence of ubiquitous comment, which becomes ever more pervasive with the spread of mobile devices. The Jotly app parodies the notion of being able to "rate everything" with a phone. With a device like Google Glass, which is a mobile computer and display built into eyeglass frames, the wearer's perspective on the world is augmented with online comment: people and locations can be decorated with informative pop-ups. Perhaps one day, people's online presence will be annotated with their media influence or dateability scores. People who wear such devices will also need new ways to comment, perhaps with hand signals. Liking something could be as simple as framing something within a heart-shaped hand gesture, as proposed in a recent patent from Google.[30] Billions of comments are posted online today, but this is just the beginning.

One question that this book asks is if we really are better off avoiding comment. Despite the fact that some sites are disabling comment, it is not easily escaped. Comment is a characteristic of contemporary life: it

COMMENTS, LIKES, VOTES,
REPLIES, RETWEETS, ...
ENJOY...

... 10 MINUTES WITHOUT!

can inform, improve, and shape people for the better, and it can alienate, manipulate, and shape people for the worse. The negatives can seem more potent than the positives, but there are many benefits to today's comment. I like to read comments (if they are well tended), love reviews (if I can trust them), enjoy constructive feedback (although it can be difficult to hear), and am delighted that some people are able to maintain a sense of humor about it. Broad polemics about the wonders and perils of technology miss the point. Comment is with us, and we must find ways to use it effectively. Can we encourage policies and technologies that are supportive of our highest ideals? Can we not become beholden to commercial interests for our interactions with others? Can we mature into a sense of self-esteem that is not predicated on flattery, but on the ability to improve and move forward? Can we learn to occasionally step away from it all? These are not rhetorical questions, though they are difficult ones.

Difficult questions sometimes are best asked of ourselves first. Just as a good community requires moderation, individuals are at their best when mindful. I used to ask students to experiment with being offline for a weekend and to reflect on the experience, which seems like an increasingly implausible task given the pervasiveness of gadgets and the connectedness of youth. In its place, I have adopted an exercise from author Howard Rheingold in which students are asked to set an intention before going online.[31] They write the intention down and set a timer: "In the next 20 minutes, I'll write a reading response for class." I ask them to reflect on what they were doing when the alarm sounds. Many students find their way to their favorite Website without completing the task, as if sleepwalking online. Although I still recommend that students try to go offline (or at least not sleep with their phones under their pillows), they now are online far more often than they are off. Consequently, we need to understand what is happening online. What insights are gleaned from sifting through the muck? Cultivate a comment community. What examples can we learn from? Beta readers. What examples should not be followed? Mean kids. What dangers are revealed beneath the silt? Manipulation. Comment is as "commenterrible" as we let it be, and it can be understood only by reading the comments at the bottom of the Web.

Notes

Chapter 1: Comment

1. John Cloud, "The YouTube Gurus," *Time*, December 25, 2006, http://www.time.com/time/printout/0,8816,1570795,00.html.

2. xxxCRAZYchannel, "Darth Vader Breathing Ten Hours," *YouTube*, June 17, 2012, http://www.youtube.com/watch?v=un8FAjXWOBY.

3. AvoidComments, "Don't Read Comments," *Twitter*, November 28, 2012, https://twitter.com/AvoidComments.

4. Marshal Mcluhan, *Counterblast* (Toronto: McClelland and Steward, 1969), 5.

5. "Facebook's Growth in the Past Year," *Facebook*, 2013, https://www.facebook.com/photo.php?fbid=10151908376831729; Doug Gross, "On Facebook, Click 'Like' Can Help Scammers," *CNN*, January 20, 2014, http://www.cnn.com/2014/01/21/tech/social-media/facebook-like-farming/index.html.

6. Robin Dunbar, *Grooming, Gossip, and the Evolution of Language* (Cambridge: Harvard University Press, 1996); Robin Dunbar, "Gossip in Evolutionary Perspective," *Review of General Psychology* 8, no. 2 (2004): 100–110.

7. Alan Dundes, "Here I Sit: A Study of American Latrinalia," *Kroeber Anthropological Society Papers* 34 (1966): 91–105; Adam Trahan, "Identity and Ideology: The Dialogic Nature of Latrinalia," *Internet Journal of Criminology* (2011): 1–9; Robert F. Goodman and Aaron Ben-Ze'ev, eds., *Good Gossip* (Lawrence: University Press of Kansas, 1994); Aaron Ben-Ze'ev, "The Vindication of Gossip," in *Good Gossip*, ed. Robert F. Goodman and Aaron Ben-Ze'ev, 11–24 (Lawrence: University Press of Kansas, 1994); Nicholas DiFonzo and Prashant Bordia, "Rumor, Gossip and Urban Legends," *Diogenes* 54, no. 19 (2007): 28, doi:10.1177/0392192107073433.

8. Dunbar, *Grooming, Gossip, and the Evolution of Language*, 71.

9. Michael Erard, "No Comments," *New York Times Magazine*, September 20, 2013, http://www.nytimes.com/2013/09/22/magazine/no-comments.html.

10. John Markoff, "An Internet Critic Who Is Not Shy about Ruffling the Big Names in High Technology," *New York Times*, April 9, 2001, http://www.nytimes

.com/2001/04/09/business/internet-critic-who-not-shy-about-ruffling-big-names
-high-technology.html.

11. Dave Winer, "The Unedited Voice of a Person," *Scripting News*, January 1, 2007, http://scripting.com/2007/01/01.html#theUneditedVoiceOfAPerson.

12. Dave Winer, "Proposal: A New Kind of Blog Comment System," *Scripting News*, August 22, 2010, http://scripting.com/stories/2010/08/22/proposalANewKindOfBlogComm.html; Dave Winer, "No Comment," *Scripting News*, February 19, 2012, http://scripting.com/stories/2012/02/19/noComment.html.

13. Mathew Ingram, "Facebook: Dave Winer Brings the Hate," *Work Blog*, October 14, 2007, http://www.mathewingram.com/work/2007/10/14/facebook-dave-winer-brings-the-hate.

14. Newshour, "*Post* Web Site Halts Comments Section," *PBS*, January 24, 2006, http://www.pbs.org/newshour/bb/media/jan-june06/post_1-24.html; Joshua Topolsky, "We're Turning Comments Off for a Bit," *Engadget*, February 2, 2010, http://www.engadget.com/2010/02/02/were-turning-comments-off-for-a-bit; Suzanne Labarre, "Why We're Shutting Off Our Comments," *Popular Science*, September 24, 2013, http://www.popsci.com/science/article/2013-09/why-were-shutting-our-comments?src=SOC&dom=tw; Rob Beschizza, "Time to Turn Off Comments for Good," *Muckrack*, February 28, 2013, http://muckrack.com/Beschizza/statuses/307172805252894722.

15. "Introducing Our Comments Box Launch Partners," *Facebook*, March 3, 2011, http://www.facebook.com/note.php?note_id=187369061298591&comments; Farhad Manjoo, "Anonymous Comments: Why We Need to Get Rid of Them Once and for All," *Anonymous Comments*, March 9, 2011, http://www.slate.com/articles/technology/technology/2011/03/troll_reveal_thyself.html.

16. MG Siegler, "Facebook Comments Have Silenced the Trolls—But Is It Too Quiet?," *TechCrunch*, March 6, 2011, http://techcrunch.com/2011/03/06/techcrunch-facebook-comments.

17. Clay Shirky, "Group as User: Flaming and the Design of Social Software," *Shirky*, November 5, 2004, http://www.shirky.com/writings/group_user.html; Clay Shirky, "'Hottubbing' as a Community Management Pattern," *Many-to-Many*, May 15, 2003, http://many.corante.com/archives/2003/05/15/hottubbing_as_a_community_management_pattern.php; Clay Shirky, *Cognitive Surplus: Creativity and Generosity in a Connected Age* (New York: Penguin Press, 2010), 197–202; Clay Shirky, "Why Do Comments Suck?," *Microsoft Research*, January 17, 2013, http://research.microsoft.com/apps/video/dl.aspx?id=180323.

18. Lawrence Lessig, "Announcing the Hibernation of Lessig.org/blog," *Blog*, August 20, 2009, http://www.lessig.org/2009/08.

19. Alice Marwick and danah boyd, "To See and Be Seen: Celebrity Practice on Twitter," *Convergence* 17, no. 2 (May 2011): 139–158.

20. Mor Naaman, Jeffrey Boase, and Chih-Hui Lai, "Is It Really about Me?," in *Proceedings of CSCW 2010*, 2010, http://infolab.stanford.edu/~mor/research/naamanCSCW10.pdf.

21. Diana I. Tamir and Jason P. Mitchell, "Disclosing Information about the Self Is Intrinsically Rewarding," *Proceedings of the National Academy of Sciences* 109 (2012): 8038–8043, doi:10.1073/pnas.1202129109.

22. Trent Reznor, "Online Communities, Etc.," *NIN Forums*, June 10, 2009, http://forum.nin.com/bb/read.php?59,731489.

23. Admin, "The Atheism+ Block Bot," *The Blockbot*, March 7, 2013, http://www.theblockbot.com.

24. Ethan Zuckerman, "The Tweetbomb and the Ethics of Attention," *My Heart's in Accra*, April 20, 2012, http://www.ethanzuckerman.com/blog/2012/04/20/the-tweetbomb-and-the-ethics-of-attention.

25. Xeni Jardin, "The Diagnosis," *Boing Boing*, December 9, 2011, http://boingboing.net/2011/12/09/the-diagnosis.html.

26. Dhiraj Murthy, *Twitter: Social Communication in the Twitter Age* (Malden, MA: Polity, 2013), 140.

27. Xeni Jardin, "Twitter / Xeni: I Have Breast Cancer. I Am ...," *Twitter*, December 1, 2011, https://twitter.com/xeni/status/142437402626105344.

28. Reese Leysen, "ShareCraft Celebrity Tweet-Bombing (and Other Ideas)," *I Power*, April 6, 2012, http://ipowerproject.com/forum/topics/sharecraft-celebrity-tweet-bombing-and-other-ideas.

29. Zuckerman, "The Tweetbomb and the Ethics of Attention."

30. mc rider, "@xeni All These People Sacrifice ...," *Twitter*, April 19, 2012, https://twitter.com/wesir23/status/193089878580400128.

31. Robert J. Nemiroff and Teresa Wilson, "Searching the Internet for Evidence of Time Travelers," *Arxiv*, December 26, 2013, http://arxiv.org/abs/1312.7128.

Chapter 2: Informed

1. Aaron Smith, "How Americans Used Their Phones to Assist with Purchasing Decisions This Holiday Season," *Pew Internet and American Life Project*, January 30, 2012, http://www.pewinternet.org/Reports/2012/In-store-mobile-commerce/Findings.aspx.

2. Barry Schwartz, *The Paradox of Choice: Why More Is Less* (New York: HarperCollins, 2004), 78, http://books.google.com/books?id=zutxr7rGc_QC.

3. Although marginal annotations can be seen on the papyrus rolls from the second and third centuries CE, scholia emerged in force in the margins of the codex (a bound book) in the late Rome and early Byzantium periods. The form and conventions of the book are now taken for granted, but the *mise-en-page*, or layout of a page, underwent significant developments during this period. The earliest codices often had four columns of text, emulating the layout of papyrus scrolls. During the late antique and early Christian periods, scribes moved toward one- or two-column layouts, making room for commentary in the margins. This practice of annotation continued in the east in the Byzantium empire and came to be known as *scholia recentiora* relative to the older *scholia vetera* of the Greeks

and Romans. As the Roman state failed, the western church used marginal annotations to parse and clarify the works of the early Christian fathers (the *patristic* texts). In fact, the early Christians often are credited with popularizing the codex (book) format because they valued its inexpensiveness and convenience. Some even hung frequently consulted texts from their waist belts. This type of pocket reference books was called a *vade mecum* (Latin for "go with me"), much like our mobile devices today. See Michelle P. Brown, *Understanding Illuminated Manuscripts: A Guide to Technical Terms* (Malibu, CA: J. Paul Getty Museum, British Library Board, 1994), 86, 96, 124.

Originally, texts and their commentary were separate works, but even narrow margins provided enough space for readers to write signs that indicated noteworthy text (using the *diple*, like a ">"), spurious text (using the *obelos*, like a "–"), or references to an external commentary. See L. D. Reynolds and N. G. Wilson, *Scribes and Scholars: A Guide to the Transmission of Greek and Latin Literature*, 3rd ed. (New York: Oxford University Press, 1991), 10–11. Conversely, commentaries used *lemmata*, short quotations indicating the word or passage from the source text under discussion. With codices and wider margins, commentary itself could be included alongside the text. Yet this too could lead to confusion as tired or lazy copyists might abbreviate, omit, miscopy, or mistakenly swap scholia with its source. Scholia are often more plentiful at the beginning of a long work than in subsequent sections, which is evidence of copyists' fatigue. See Eleanor Dickey, *Ancient Greek Scholarship: A Guide to Finding, Reading, and Understanding Scholia, Commentaries, Lexica, Grammatical Treatises, from Their Beginnings to the Byzantine Period* (New York: Oxford University Press, 2007), 12, 14.

4. Tom Standage, *Writing on the Wall: Social Media—the First Two Thousand Years* (New York: Bloomsbury, 2013); Pope Leo X, "Exsurge Domine: Condemning the Errors of Martin Luther," *EWTN*, June 15, 1520, http://www.ewtn.com/library/papaldoc/l10exdom.htm.

5. Gottfried Leibniz, 1680, quoted in W. Boyd Rayward, "Some Schemes for Restructuring and Mobilizing Information in Documents: A Historical Perspective," *Information Processing and Management* 30, no. 2 (1994): 167.

6. Coetlogon, quoted in Jeff Loveland, *An Alternative Encyclopedia? Dennis de Coetlogon's Universal History* (1745), Studies on Voltaire and the Eighteenth Century (Oxford: Voltaire Foundation, University of Oxford, 2010), 71.

7. Jürgen Habermas, *The Structural Transformation of the Public Sphere: An Inquiry into a Category of Bourgeois Society*, trans. Thomas Burger and Frederick Lawrence (Cambridge, MA: MIT Press, 1989), 23, 59, http://pages.uoregon.edu/koopman/courses_readings/phil123-net/publicness/habermas_structural_trans_pub_sphere.pdf.

8. "Advertisement," *Monthly Review*, 1749, quoted in Frank Donoghue, *The Fame Machine: Book Reviewing and Eighteenth-Century Literary Careers* (Stanford, CA: Stanford University, 1996), 23.

9. "Book Review One Hundred Years," *New York Times*, October 6, 1996, http://www.nytimes.com/1996/10/06/books/book-review-100-years.html; Gabriel Compayré and W. H. Payne, *History of Pedagogy* (Boston: Cushing, 1885), 571–572;

Brian N. Cary, "Gunning for the American Film Company: Wid Gunning and Wid's Film and Film Folk," *UCSB*, 2013, http://db.oic.id.ucsb.edu:8090/4DACTION/www_ShowEssayDetail?essayID=6.

10. David Weinberger, *Everything Is Miscellaneous: The Power of the New Digital Disorder* (New York: Times Books, 2007).

11. J Street, "Mr. Bacon's 2.5 Oz Bacon-Flavored Toothpaste," *Amazon*, January 7, 2014, http://www.amazon.com/review/RHROHSQCATCDI/ref=cm_cr_rdp_perm?ie=UTF8&ASIN=B004MBNK5K&linkCode=&nodeID=&tag=.

12. MG Siegler, "Facebook: We'll Serve One Billion Likes on the Web in Just Twenty-four Hours," *TechCrunch*, April 21, 2010, http://techcrunch.com/2010/04/21/facebook-like-button; Sandra Liu Huang, "After F8—Resources for Building the Personalized Web," *Facebook Developers*, April 28, 2010, http://developers.facebook.com/blog/post/2010/04/28/after-f8---resources-for-building-the-personalized-web.

13. "Full of Fire," *The Knife*, January 29, 2013, http://theknife.net/full-of-fire; Add This, "The Largest Sharing and Social Data Platform. We Provide Twitter and Facebook Buttons, Custom Audience Targeting, and More," *Addthis*, January 29, 2013, http://www.addthis.com; Oliver Reichenstein, "Sweep the Sleaze," *Information Architects*, May 29, 2012, http://informationarchitects.net/blog/sweep-the-sleaze.

14. Herbert Lottman, *The Michelin Men: Driving an Empire* (New York: Tauris, 2003), 2–4, 146–148.

15. Ibid., 44, 49, 78.

16. Chase, quoted in Robert B. Westbrook, "Tribune of the Technostructure: The Popular Economics of Stuart Chase," *American Quarterly* 32, no. 4 (1980): 387–408.

17. Stuart Chase and F.J. Schlink, *Your Money's Worth: A Study in the Waste of the Consumer's Dollar*, *Library of Congress* (New York: Macmillan, 1927), 2, http://memory.loc.gov/cgi-bin/query/r?ammem/cool:@field(DOCID+@lit(lg07T000))::bibLink=r?ammem/coolbib%3A@FIELD(SUBJ+@band(+Testing.+)).

18. Ibid., 5, 11.

19. Arthur Kallet and F. J. Schlink, *One Hundred Million Guinea Pigs: Dangers in Everyday Foods, Drugs, and Cosmetics* (New York: Vanguard Press, 1932), 4, 209.

20. Consumers Union, "Our History," *ConsumerReports*, September 20, 2004, 1935, http://www.consumerreports.org/cro/aboutus/history/printable/index.htm.

21. Consumers Union, "Consumers Union Shopping and Testing," *ConsumerReports*, June 13, 2006, http://web.archive.org/web/20060616222830/http://www.consumerreports.org/cro/cu-press-room/pressroom/shoppingtesting/index.htm.

22. Kevin Kelly, "Cool Tools," *Kk*, January 30, 2013, http://kk.org/cooltools.

23. Stewart Brand, "Photography Changes Our Relationship to Our Planet," *Smithsonian Photography Initiative*, 2010, http://click.si.edu/Story.aspx?story=31.

24. Stewart Brand, "The Purpose," *Whole Earth Catalog*, 1968, http://www
.wholeearth.com/issue/1010/article/196/the.purpose.of.the.whole.earth.catalog;
Stewart Brand, "Introduction to Whole Earth Software Catalog," *Whole Earth
Catalog*, 1984, http://www.wholeearth.com/issue/1230/article/283/introduction
.to.whole.earth.software.catalog.

25. Stewart Brand, "We Owe It All to the Hippies," *Time* 145, no. 12 (1995),
http://members.aye.net/~hippie/hippie/special_.htm.

26. Fred Turner, "Where the Counterculture Met the New Economy: The WELL
and the Origins of Virtual Community," *Technology and Culture* 46, no. 3 (2005):
507–510.

27. Mark Frauenfelder, "Mark Joins Cool Tools," *Boing Boing*, February 4, 2013,
http://boingboing.net/2013/02/04/mark-joins-cool-tools.html; Kyle Fitzpatrick,
"The Culture Maker: An Interview with Mark Frauenfelder of Boing Boing," *Los
Angeles, I'm Yours*, May 21, 2012, http://www.laimyours.com/17922/the-culture
-maker-an-interview-with-mark-frauenfelder-of-boing-boing.

28. Kevin Kelly, "Introducing a Catalog of Possibilities: Cool Tools," *YouTube*,
November 8, 2013, http://www.youtube.com/watch?v=332hGlHLfMA.

29. "Veggie Galaxy," *Zagat*, March 24, 2014, http://www.zagat.com/r/veggie
-galaxy-cambridge.

30. Marissa Mayer, "Google Just Got ZAGAT Rated!," *Google Official Blog*,
September 8, 2011, http://googleblog.blogspot.com/2011/09/google-just-got
-zagat-rated.html.

31. Tim O'Reilly, "Web 2.0 Compact Definition: Trying Again," December 10,
2006, http://radar.oreilly.com/2006/12/web-20-compact-definition-tryi.html.

32. Kevin Kelly, *Out of Control: The New Biology of Machines, Social Systems
and the Economic World* (Cambridge, MA: Perseus Books Group, 1995), http://
www.kk.org/outofcontrol.

33. Howard Rheingold, *Smart Mobs: The Next Social Revolution* (Cambridge,
MA: Perseus, 2002).

34. James Surowiecki, *The Wisdom of Crowds: Why the Many Are Smarter Than
the Few and How Collective Wisdom Shapes Business, Economies, Societies, and
Nations* (New York: Doubleday, 2004).

35. Los Angeles Times, "In Online World, Everyone Can Be a Critic," *Sun Sentinel*,
June 17, 1997, http://articles.sun-sentinel.com/1997-06-17/lifestyle/9706160115
_1_internet-life-review-http.

36. James Berardinelli, "Reel Thoughts," *Reelviews*, 2012, http://reelviews.net/
faq.html.

37. James Berardinelli and Dan Schneider, "Interview 16," *Cosmoetica*, December 8, 2012, http://www.cosmoetica.com/DSI16.htm.

38. Stephen Sondheim, "Critics and Their Uses," in *Look, I Made a Hat: Collected Lyrics (1981–2011) with Attendant Comments, Amplifications, Dogmas,
Harangues, Digressions, Anecdotes, and Miscellany* (New York: Knopf, 2011),
40.

39. Berardinelli, "Reel Thoughts."

40. Jason Silverman, "Invasion of the Web Film Critics," *Wired*, February 28, 2004, http://archive.wired.com/entertainment/music/news/2004/02/62453?currentPage =all.

41. Jennifer McDonald, "Masters of the Form," *New York Times*, December 31, 2010, http://www.nytimes.com/2011/01/02/books/review/LitCritBackPage-t .html; Editors, "Up Front: Why Criticism Matters," *New York Times*, December 31, 2010, http://www.nytimes.com/2011/01/02/books/review/Tanenhaus-t.html ?_r=1.

42. Stephen Burn, "Beyond the Critic as Cultural Arbiter," *New York Times*, December 31, 2010, http://www.nytimes.com/2011/01/02/books/review/Burn-t -web.html; Sam Anderson, "Translating the Code into Everyday Language," *New York Times*, December 31, 2010, http://www.nytimes.com/2011/01/02/books/ review/Anderson-t-web.html; Katie Roiphe, "With Clarity and Beauty, the Weight of Authority," *New York Times*, December 31, 2010, http://www.nytimes.com/ 2011/01/02/books/review/Roiphe-t-web.html.

43. Vint Cerf, "The Future of the Internet," *The European*, September 27, 2012, http://www.theeuropean-magazine.com/vint-cerf/834-the-future-of-the-internet.

44. "Unboxing," *Wikipedia*, February 5, 2013, http://en.wikipedia.org/?oldid =536740195; Heather Kelly, "The Bizarre, Lucrative World of 'Unboxing' Videos," *CNN*, February 12, 2014, http://www.cnn.com/2014/02/13/tech/web/ youtube-unboxing-videos/index.html.

45. hipstomp / Rain Noe, "Why Survivalists Make Great Bag Reviewers," *Core77*, August 9, 2012, http://www.core77.com/blog/cases/why_survivalists_make_great _bag_reviewers_23043.asp.

46. mokyan7, "Maxpedition Pygmy Falcon II: A Review," *YouTube*, October 11, 2011, http://www.youtube.com/watch?v=AwiwIJlLGYc.

47. Kodiakbear, "Kodiakbear's Review of Maxpedition Pygmy Falcon-II Backpack (Khaki)," *Amazon*, September 2, 2011, http://www.amazon.com/review/ R307427YGQ1AIN/ref=cm_cr_pr_perm?ie=UTF8&ASIN=B001E8EXHS &linkCode=&nodeID=&tag=.

48. Todd B., "Todd B.'s Review of Maxpedition Pygmy Falcon-II," *Amazon*, November 15, 2011, http://www.amazon.com/review/R3UHV2MVDF6NKM/ ref=cm_cr_rdp_perm?ie=UTF8&ASIN=B004Q5E6YG&linkCode=&nodeID= &tag=.

Chapter 3: Manipulated

1. Alex Cornell, "Jotly: Rate Everything," *YouTube*, October 13, 2011, http:// www.youtube.com/watch?v=QIWpbfZHHzc.

2. "Jotly: Rate Everything," *Firespotter Labs*, May 23, 2013, http://www.jotly.co.

3. Alex Cocotas, "How $3 Billion TripAdvisor's Business Works," *Business Insider*, December 23, 2011, http://www.businessinsider.com/how-tripadvisors

-business-works-2011-12; Erick Schonfeld, "Google Places Now Borrowing Yelp Reviews without Attribution in IPhone App," *TechCrunch*, June 1, 2011, http://techcrunch.com/2011/06/01/google-places-borrowing-yelp-iphone-app; Jason Kincaid, "Google Acquires Zagat to Flesh Out Local Reviews," *TechCrunch*, September 8, 2011, http://techcrunch.com/2011/09/08/google-acquires-zagat-to-flesh-out-local-ratings; Megan Geuss, "Google Mines Frommer's Travel for Social Data, Then Sells the Name Back," *Ars Technica*, April 9, 2013, http://arstechnica.com/business/2013/04/google-mines-frommers-travel-for-social-data-then-sells-the-name-back.

4. Jeff Bezos, quoted in Robert Spector, *Amazon.com: Get Big Fast* (New York: HarperCollins, 2000), 132.

5. Jim Jansen, "58 Percent of Americans Have Researched a Product or Service Online," *Pew Internet & American Life Project*, September 29, 2010, http://www.pewinternet.org/2010/09/29/online-product-research; Lee Rainie et al., "Where People Get Information about Restaurants and Other Local Businesses," *Pew Internet & American Life Project*, December 14, 2011, http://www.pewinternet.org/Reports/2011/Local-business-info/Overview.aspx?view=all.

6. Paul Resnick et al., "The Value of Reputation on EBay: A Controlled Experiment," *Experimental Economics* 9, no. 2 (2006): 79–101; Michelle Haynes and Steve Thompson, "The Economic Significance of User-Generated Feedback," *International Journal of the Economics of Business* 19, no. 1 (2012): 153–166, doi:10.1080/13571516.2012.642645; Judith A Chevalier and Dina Mayzlin, "The Effect of Word of Mouth on Sales: Online Book Reviews," *Journal of Marketing Research* 43, no. 3 (2006): 345–54, doi:10.1509/jmkr.43.3.345.

7. Michael Luca, "Reviews, Reputation, and Revenue: The Case of Yelp.com," *Harvard Business School Working Paper*, September 16, 2011, http://ctct.wpengine.com/wp-content/uploads/2011/10/12-016.pdf; Michael Anderson and Jeremy Magruder, "Learning from the Crowd: Regression Discontinuity Estimates of the Effects of an Online Review Database," *Economic Journal* 122, no. 563 (September 2012): 957–989, doi:10.1111/j.1468-0297.2012.02512.x; Hulisi Öğüt and Bedri Kamil Onur Taş, "The Influence of Internet Customer Reviews on the Online Sales and Prices in Hotel Industry," *Service Industries Journal* 32, no. 2 (2012): 197–214, doi:10.1080/02642069.2010.529436; Feng Zhu and Xiaoquan (Michael) Zhang, "Impact of Online Consumer Reviews on Sales: The Moderating Role of Product and Consumer Characteristics," *Journal of Marketing* 74, no. 2 (2010): 133–148, doi:10.1509/jmkg.74.2.133; Wenjing Duan, Bin Gu, and Andrew B. Whinston, "Do Online Reviews Matter? An Empirical Investigation of Panel Data," *Decision Support Systems* 45, no. 4 (November 2008): 1007–1016.

8. Peter De Maeyer, "Impact of Online Consumer Reviews on Sales and Price Strategies: A Review and Directions for Future Research," *Journal of Product & Brand Management* 21, no. 2 (2012): 132–139.

9. Shay David and Trevor Pinch, "Six Degrees of Reputation: The Use and Abuse of Online Review and Recommendation Systems," *First Monday*, July 2006, http://firstmonday.org/htbin/cgiwrap/bin/ojs/index.php/fm/article/view/1590/1505.

10. Nitin Jindal and Bing Liu, "Opinion Spam and Analysis," in *Proceedings of First ACM International Conference on Web Search and Data Mining (WSDM 2008)* (New York: ACM, 2008), http://www.cs.uic.edu/~liub/FBS/opinion-spam -WSDM-08.pdf; David Streitfeld, "The Best Book Reviews Money Can Buy," *New York Times*, August 25, 2012, http://www.nytimes.com/2012/08/26/business/ book-reviewers-for-hire-meet-a-demand-for-online-raves.html.

11. Nan Hu et al., "Manipulation of Online Reviews: An Analysis of Ratings, Readability, and Sentiments," *Decision Support Systems* 52, no. 3 (2012): 681, doi:10.1016/j.dss.2011.11.002.

12. Michael Luca and Georgios Zervas, "Fake It Till You Make It: Reputation, Competition, and Yelp Review Fraud," 2013, http://papers.ssrn.com/sol3/papers .cfm?abstract_id=2293164.

13. Donna Gibbs, "The Language of Cyberspace," in *Cyberlines 2.0: Languages and Cultures of the Internet*, ed. Donna Gibbs and Kerri-Lee Krause (Albert Park, Australia: James Nicholas, 2006), 30.

14. Mathias Persson, "A Veil of Ignorance: Anonymity and Promotion of Self in the Eighteenth-Century Republic of Letters," *Proceedings of The Emergence of the Periodical Form (Seventeenth–Eighteenth Centuries) as an Instrument of Scientific Change (Symposium S-42, ICHST09)*, Budapest, 2009; Bob Fenster, *The Duh Awards: In This Stupid World, We Take the Prize* (Kansas City: Andrews McMeel, 2005).

15. Walter Scott, "Living Poets of Great Britain," *Edinburgh Annual Register*, 1808, 423–426.

16. David Haven Blake, *Walt Whitman and the Culture of American Celebrity* (New Haven, CT: Yale University Press, 2006), 114.

17. Anthony Burgess, "Anthony Burgess: Confessions of the Hack Trade," *The Observer*, March 3, 2012, http://www.guardian.co.uk/culture/2012/mar/04/ anthony-burgess-on-journalism-1992.

18. John Rechy, quoted in Amy Harmon, "Amazon Glitch Unmasks War of Reviewers," *New York Times*, February 14, 2004, http://www.nytimes.com/2004/ 02/14/us/amazon-glitch-unmasks-war-of-reviewers.html.

19. Richard Lea and Matthew Taylor, "Historian Orlando Figes Admits Posting Amazon Reviews That Trashed Rivals," *The Guardian*, April 23, 2010, http://www.guardian.co.uk/books/2010/apr/23/historian-orlando-figes-amazon -reviews-rivals.

20. Diane Purkiss, "Two Out of Five Stars: Too Jolly by Half," *Amazon*, April 8, 2010, http://www.amazon.co.uk/review/R3QV4JC3CIU7BI.

21. Ibid.

22. Scott Adams, "I'm a What?," *Scott Adams Blog*, March 27, 2011, http://www .dilbert.com/blog/entry/im_a_what; Scott Adams, "How to Get a Real Education," *Metafilter*, April 12, 2011, http://www.metafilter.com/102472/How-to-Get -a-Real-Education-by-Scott-Adams#3639589.

23. Johann Hari, "A Personal Apology," *The Independent*, September 15, 2011, http://www.independent.co.uk/voices/commentators/johann-hari/johann-hari-a -personal-apology-2354679.html.

24. David Streitfeld, "A Casualty on the Battlefield of Amazon's Partisan Book Reviews," *New York Times*, January 20, 2013, http://www.nytimes.com/2013/ 01/21/business/a-casualty-on-the-battlefield-of-amazons-partisan-book-reviews .html?_r=0; Padmananda Rama, "Yelp Reviewers Slice and Dice the Politics of Pizza," *NPR: The Salt*, September 11, 2012, http://www.npr.org/blogs/thesalt/ 2012/09/11/160960730/yelp-reviewers-slice-and-dice-the-politics-of-pizza?ft=1 &f=1001; Eric Anderson and Duncan Simester, "Deceptive Reviews: The Influential Tail," May 2013, http://web.mit.edu/simester/Public/Papers/Deceptive _Reviews.pdf.

25. Shawn Knight, "Samsung Admits to Posting Fake User Reviews on the Web," *Techspot*, April 17, 2013, http://www.techspot.com/news/52274-samsung-admits -to-posting-fake-user-reviews-on-the-web.html; Rob Marvin, "Samsung Buying Off StackOverflow Users for Publicity," *Software Development News*, August 1, 2013, http://sdt.bz/61968; Happy Rockefeller, "The HB Gary Email That Should Concern Us All," *Daily Kos*, February 16, 2011, http://www.dailykos.com/story/ 2011/02/16/945768/-UPDATED-The-HB-Gary-Email-That-Should-Concern-Us -All; Michael Bristow, "China's Internet 'Spin Doctors,'" *BBC News*, December 16, 2008, http://news.bbc.co.uk/2/hi/asia-pacific/7783640.stm; Sergey Chernov, "Internet Troll Operation Uncovered in St. Petersburg," *St. Petersburg Times*, September 18, 2013, http://www.sptimes.ru/index.php?action_id=100&story _id=38052; Prachatai, "Cyber Soldiers Promote the Monarchy," *Prachatai English*, July 6, 2013, http://www.prachatai.com/english/node/3617; Choe Sang-Hun, "Prosecutors Detail Attempt to Sway South Korean Election," *New York Times*, November 21, 2013, http://www.nytimes.com/2013/11/22/world/asia/ prosecutors-detail-bid-to-sway-south-korean-election.html.

26. Graeme Wood, "The World of Black-Ops Reputation Management," *New York Magazine*, June 16, 2013, http://nymag.com/news/features/online-reputation -management-2013-6.

27. Daily Mail Reporter, "Amazon and TripAdvisor at Centre of Scandal as Companies Post Fake Reviews," *Mail Online*, June 2, 2011, http://www.dailymail.co .uk/news/article-1393412/Amazon-TripAdvisor-centre-scandal-companies-post -fake-reviews.html?ito=feeds-newsxml; Daniela Hernandez, "Homeless, Unemployed, and Surviving on Bitcoins," *Wired*, September 20, 2013, http://www .wired.com/2013/09/bitcoin-homeless/all; "Over One Hundred Fifteen BTC Paid Out in Six Months," *BitCoinGet*, August 8, 2013, http://www.bitcoinget.com/ blog/over-115-btc-paid-out-in-six-months.

28. Mutaaly quoted in Mike Deri Smith, "Fake Reviews Plague Consumer Websites," *The Guardian*, January 25, 2013, http://www.guardian.co.uk/money/2013/ jan/26/fake-reviews-plague-consumer-websites.

29. Garth Hallberg, "Who Is Grady Harp? Amazon's Top Reviewers and the Fate of the Literary Amateur," *Slate*, January 22, 2008, http://www.slate.com/articles/ arts/culturebox/2008/01/who_is_grady_harp.html.

30. Joshua Porter, "Is Harriet Klausner for Real?," *Bokardo*, October 15, 2007, http://bokardo.com/archives/is-harriet-klausner-for-real; Sneaky Burrito, "She Works Hard for the Money," The Harriet Klausner Appreciation Society, October 9, 2012, http://harriet-rules.blogspot.com/2012/10/she-works-hard-for-money .html.

31. Lee Goldberg, "Review of How I Sold One Million EBooks in Five Months," *Amazon*, August 26, 2012, http://www.amazon.com/review/ R1OO6OYAZKYLTR/ref=cm_cr_pr_viewpnt#R1OO6OYAZKYLTR.

32. "Author Services FAQ," *Kirkus*, August 11, 2012, https://www.kirkusreviews .com/author-services/indie/faq.

33. Kurt Thomas et al., "Trafficking Fraudulent Accounts: The Role of the Underground Market in Twitter Spam and Abuse," in *Proceedings of the Twenty-second USENIX Security Symposium* (2013), 58.

34. Joseph Reagle, "Revenge Rating and Tweak Critique at Photo.net," in *Evaluating Creative Production in Digital Environments* (New York: Routledge, 2014), http://reagle.org/joseph/2013/photo/photo-net.html; Marilyn Strathern, "'Improving Ratings': Audit in the British University System," *European Review* 5, no. 3 (July 1997): 308.

35. Jason Ding, "The Twitter Underground Economy: A Blooming Business," *Barracuda Labs*, May 14, 2012, http://www.barracudalabs.com/wordpress/?p =2989; Mat Honan, "How to Use Social Media to Juice Your Story's Popularity," *Wired*, July 16, 2013, http://www.wired.com/gadgetlab/2013/07/cheat-page; Jim Finkle, "Virus Targets the Social Network in New Fraud Twist," *Reuters*, August 16, 2013, http://www.reuters.com/article/2013/08/16/us-instagram-cyberfraud -idUSBRE97F0XD20130816.

36. "Amazon's Top Customer Reviewers," *Amazon*, June 11, 2012, http://www .amazon.com/review/guidelines/top-reviewers.html.

37. Elie Bursztein et al., "How Good Are Humans at Solving CAPTCHAS? A Large-Scale Evaluation," in *Proceedings of IEEE Symposium on Security and Privacy* (Washington, DC: IEEE Computer Society, 2010), 399–400, doi:10.1109/ SP.2010.31.

38. Harry Brignull, "F**K CAPTCHA," *Ninety Percent of Everything*, March 25, 2011, http://www.90percentofeverything.com/2011/03/25/fk-captcha.

39. Amazon, "Welcome," *Amazon Mechanical Turk*, June 12, 2013, https://www .mturk.com/mturk/welcome.

40. Panos Ipeirotis, Dahn Tamir, and Priya Kanth, "Mechanical Turk: Now with 40.92 Percent Spam," *A Computer Scientist in a Business School*, December 16, 2010, http://www.behind-the-enemy-lines.com/2010/12/mechanical-turk-now -with-4092-spam.html.

41. Brandon Woodson, "The Dark Side of Amazon's Mechanical Turk," *Yahoo Voices*, May 2, 2010, http://voices.yahoo.com/the-dark-side-amazons-mechanical -turk-5918276.html.

42. "Cheapest CAPTCHA Bypass Service," *Death by CAPTCHA*, June 11, 2013, http://www.deathbycaptcha.com/user/login; "Death by Captcha," *Death*

by CAPTCHA, October 4, 2009, http://web.archive.org/web/20101129024020/ http://www.deathbycaptcha.com/user/login; "Contact Us," *Bypass Captcha*, June 12, 2013, http://bypasscaptcha.com/contact.php.

43. SolveMedia, "Advertiser Suite: Guaranteed Engagement and Effectiveness," *Solve Media*, September 11, 2013, http://solvemedia.com/advertisers.

44. Salsa1234, "Angie's List: Buyer Beware of Billing Practice," *Pissed Consumer*, April 2, 2011, http://www.pissedconsumer.com/reviews-by-company/angies-list/ angie-s-list-buyer-beware-of-billing-practice-20110402230254.html; David, "Six Reasons Why Angie's List Sucks?," *The Upboom Blog*, July 15, 2012, http://blog .upboom.com/2010/12/28/6-reasons-why-angies-list-sucks-continue-review-on -angieslist-com/#comment-383.

45. David Streitfeld, "For $2 a Star, a Retailer Gets Five-Star Reviews," *New York Times*, January 26, 2012, http://www.nytimes.com/2012/01/27/technology/for-2 -a-star-a-retailer-gets-5-star-reviews.html.

46. "Guides Concerning the Use of Endorsements and Testimonials in Advertising (16 CFR Part 255)," Federal Trade Commission, October 9, 2009, http://web.archive.org/web/20130509030306/http://ftc.gov/os/2009/10/ 091005revisedendorsementguides.pdf; "FTC Staff Revises Online Advertising Disclosure Guidelines," Federal Trade Commission, March 12, 2013, http://www .ftc.gov/news-events/press-releases/2013/03/ftc-staff-revises-online-advertising -disclosure-guidelines.

47. A. S. Radin, "RipOffReport.com Is a Scam," *Udog*, June 13, 2013, http:// www.udog.net/ror.html; Rob Beschizza, "'Potential Prostitutes' Site Lets Users Label Women as Prostitutes, Charges 'Removal' Fees," *Boing Boing*, December 27, 2012, http://boingboing.net/2012/12/27/potential-prostitutes-site.html; Press Release, "STD Registry Makes Surfing for STDs Feel Like Shopping," 24-7 Press Release, February 13, 2012, http://www.24-7pressrelease.com/press-release/std -registry-makes-surfing-for-stds-feel-like-shopping-262420.php.

48. U.S. Congress, "Protection for Private Blocking and Screening of Offensive Material (47 USC §230)," Legal Information Institute, 1996, http://www.law .cornell.edu/uscode/text/47/230.

49. Daniel J. Solove, "Speech, Privacy, and Reputation on the Internet," in *The Offensive Internet*, ed. Saul Levmore and Martha C. Nussbaum (Cambridge, MA: Harvard University, 2010); Danielle Keats Citron, *Hate Crimes in Cyberspace* (Cambridge, MA: Harvard University Press, 2014); John W. Dozier Jr. and Sue Scheff, *Google Bomb: The Untold Story of the $11.3M Verdict That Changed the Way We Use the Internet* (Deerfield Beach, FL: Health Communications, 2009); Eric Goldman, "Should TheDirty Website Be Liable for Encouraging Users to Gossip?," *Technology & Marketing Law Blog*, December 9, 2013, http:/ /blog.ericgoldman.org/archives/2013/12/should-thedirty-website-be-liable-for -encouraging-users-to-gossip-forbes-cross-post.htm; Eric P. Robinson, "'Dirty' Verdict Sets up Section 230 Appeal," *Digital Media Law Project*, April 3, 2014, http://www.dmlp.org/blog/2013/dirty-verdict-sets-section-230-appeal;   James Lasdun, *Give Me Everything You Have: On Being Stalked* (New York: Farrar, Straus, and Giroux, 2013), 113.

50. Pissed Consumer, "Business Solutions," *Pissedconsumer*, June 13, 2013, http://www.pissedconsumer.com/business-solutions.html; "Ripoff Report Verified," *Ripoff Report*, June 19, 2013, https://verified.ripoffreport.com.

51. "VIP Arbitration Program," *Ripoff Report*, July 27, 2010, http://www.ripoffreport.com/r/Ripoff-Report-VIP-Arbitration-Program-Remove-Rip-off-Report-Better-yet-Ripoff-Report-VIP-Arbitration-Program-Reputation-Repair-Reputation-Management-services-cant-deliver/Tempe-Internet-Arizona-85280/Ripoff-Report-VIP-Arbitration-Program-Remove-Rip-off-Report-Better-yet-Ripoff-Report-VI-626838.

52. Kathleen Richards, "Yelp and the Business of Extortion 2.0," *East Bay Express*, February 18, 2009, http://www.eastbayexpress.com/ebx/yelp-and-the-business-of-extortion-20/Content?oid=1176635&showFullText=true; Kathleen Richards, "Yelp Extortion Allegations Stack up," *East Bay Express*, March 18, 2009, http://www.eastbayexpress.com/ebx/yelp-extortion-allegations-stack-up/Content?oid=1176984&showFullText=true.

53. Edward M. Chen, "Levitt V Yelp Dismissal," *Scribd*, October 14, 2011, http://www.scribd.com/doc/70421921/Levitt-v-Yelp-Dismissal; Eric Goldman, "Levitt V. Yelp: Yelp Gets Complete Win in Advertiser "Extortion" Case," *Technology & Marketing Law Blog*, October 26, 2011, http://blog.ericgoldman.org/archives/2011/10/yelp_gets_compl.htm.

54. Daniel J. Solove, *The Future of Reputation: Gossip, Rumor, and Privacy on the Internet* (New Haven, CT: Yale University Press, 2007), http://docs.law.gwu.edu/facweb/dsolove/Future-of-Reputation/text.htm; Citron, *Hate 3.0*; Ann Bartow, "Internet Defamation as Profit Center: A Monetization of Online Harassment," *Harvard Journal of Law and Gender* 32, no. 2 (2009): 429, http://www.law.harvard.edu/students/orgs/jlg/vol322/383-430.pdf.

55. Joseph Rhee, "L.A. Better Business Bureau Chapter Head Bill Mitchell Quits Job," *ABC News*, December 22, 2010, http://abcnews.go.com/Blotter/la-business-bureau-chapter-head-bill-mitchell-quits/story?id=12458713&page=1#.UbHKpnVDs44.

56. Quentyn Kennemer, "Play Store Website Gets Roboto Font, and All Developers Can Now Respond to Individual Reviews," *Phandroid*, January 10, 2013, http://phandroid.com/2013/01/10/google-play-store-developer-comments.

57. "ReviewerCard," *Reviewer Card*, June 17, 2013, http://reviewercard.com.

58. Stephanie Strom, "When 'Liking' a Brand Online Voids the Right to Sue," *New York Times*, April 16, 2014, http://www.nytimes.com/2014/04/17/business/when-liking-a-brand-online-voids-the-right-to-sue.html.

59. Justin Brookman, "CDT Files FTC Complaint against Medical Justice," *CDT*, November 29, 2011, https://cdt.org/blog/cdt-files-ftc-complaint-against-medical-justice; Joe Mullin, "Dentist Who Used Copyright to Silence Her Patients Is on the Run," *Ars Technica*, July 28, 2013, http://arstechnica.com/tech-policy/2013/07/dentist-who-used-copyright-to-silence-her-patients-is-on-the-run.

60. Scott Michelman, "Utah Couple Seeks Relief after Credit Ruined over "Non-Disparagement" Clause in a Website's Fine," *Consumer Law and Privacy Blog*,

November 25, 2013, http://pubcit.typepad.com/clpblog/2013/11/utah-couple
-seeks-relief-after-credit-ruined-over-non-disparagement-clause-in-a-websites
-fine-print.html.

61. Ryan Tate, "Yelp Fights Make Leap to Real-World Violence, Says Reviewer,"
Gawker, November 3, 2009, http://gawker.com/5396122/yelp-fights-make-leap
-to-real+world-violence-says-reviewer; David Streitfeld, "Amazon Book Reviews
Deleted in a Purge Aimed at Manipulation," *New York Times*, December 22,
2012, http://www.nytimes.com/2012/12/23/technology/amazon-book-reviews
-deleted-in-a-purge-aimed-at-manipulation.html; Doug Gross, "Facebook Cracking Down on Fake 'Likes,'" *CNN*, September 27, 2012, http://www.cnn.com/
2012/09/27/tech/social-media/facebook-fake-likes/index.html; Chase Hoffberger,
"YouTube Strips Universal and Sony of Two Billion Fake Views," *Daily Dot*,
December 21, 2012, http://www.dailydot.com/news/youtube-universal-sony-fake
-views-black-hat; Michael Learmonth, "As Fake Reviews Rise, Yelp, Others Crack
Down on Fraudsters," *Advertising Age*, October 30, 2012, http://adage.com/
article/digital/fake-reviews-rise-yelp-crack-fraudsters/237486.

62. George Packer, "Is Amazon Bad for Books?," *The New Yorker*, February 17,
2014, http://www.newyorker.com/reporting/2014/02/17/140217fa_fact_packer
?currentPage=all; "Promote the Products You Sell on Amazon.com with Keyword
Targeted Ads," *Amazon*, January 1, 2012, http://services.amazon.com/services/
sponsored-products-questions.htm.

63. jasax, "Are Amazon Vine Reviews of Technical Books a Joke?," *Slashdot*,
July 13, 2013, http://news.slashdot.org/story/13/07/13/033228/are-amazon-vine
-reviews-of-technical-books-a-joke.

64. "What Are Sponsored Stories?," *Facebook*, June 19, 2013, https://www
.facebook.com/help/162317430499238.

65. Passion Lubes, "Passion Natural Water-Based Lubricant: Fifty-five Gallon,"
Amazon, October 2, 2011, http://www.amazon.com/Passion-Natural-Water
-Based-Lubricant-Gallon/dp/B005MR3IVO; George Takei, "George Takei's
Review of Passion Natural Water-Based Lubricant," *Amazon*, June 15, 2013,
http://www.amazon.com/review/R3IKIXCIYNNWVD/ref=cm_cr_pr_perm?ie
=UTF8&ASIN=B005MR3IVO&linkCode=&nodeID=&tag=; Somini Sengupta,
"On Facebook, 'Likes' Become Ads," *New York Times*, May 31, 2012, http://
www.nytimes.com/2012/06/01/technology/so-much-for-sharing-his-like.html.

66. Josh Constine, "Facebook Sponsored Stories Ads Have 46 Percent Higher
CTR, 18 Percent Lower Cost per Fan Says TBG Digital Test," *Inside Facebook*,
May 3, 2011, http://www.insidefacebook.com/2011/05/03/sponsored-stories-ctr
-cost-per-fa; Lauren Indvik, "Facebook: Sponsored Stories Make $1 Million a
Day," *Mashable*, July 26, 2012, http://mashable.com/2012/07/26/facebook-q2
-2012-earnings-call; *Fraley v. Facebook*, Fraley Facebook Settlement, May 2,
2013, http://fraleyfacebooksettlement.com/index; Cotton Delo, "Facebook Drops
'Sponsored Stories' as It Cuts Ad Formats," *Advertising Age*, June 6, 2013, http:/
/adage.com/article/digital/facebook-drops-sponsored-stories-cuts-ad-formats/
241969.

67. David Fleck, "Heads Up: We're Testing a New Form of Advertising," *Disqus: The Official Blog*, April 7, 2014, http://blog.disqus.com/post/82003625662/heads-up-were-testing-a-new-form-of-advertising.

68. "Terms of Service Update: Policies and Principles," *Google*, November 11, 2013, http://www.google.com/policies/terms/changes.

69. Ashish Bhatia, "Automated Generation of Suggestions for Personalized Reactions in a Social Network," United States Patent Office, November 19, 2013, http://patft.uspto.gov/netacgi/nph-Parser?Sect1=PTO1&Sect2=HITOFF&d=PALL&p=1&u=%2Fnetahtml%2FPTO%2Fsrchnum.htm&r=1&f=G&l=50&s1=8,589,407.PN.&OS=PN/8,589,407&RS=PN/8,589,407.

70. Derek Muller, "Facebook Fraud," *YouTube*, February 10, 2014, http://www.youtube.com/watch?v=oVfHeWTKjag; Muller, quoted in Andrew Leonard, "Facebook's Black Market Problem Revealed," *Salon*, February 14, 2014, http://www.salon.com/2014/02/14/facebooks_big_like_problem_major_money_and_major_scams; Jim Edwards, "This Man's $600,000 Facebook Disaster Is a Warning for All Small Businesses (FB)," *SFGate*, February 27, 2014, http://www.sfgate.com/technology/businessinsider/article/This-Man-s-600-000-Facebook-Ad-Disaster-Is-A-5258472.php.

71. Oliver Widder, "The 'Free' Model," *Geek and Poke*, December 21, 2010, http://geekandpoke.typepad.com/geekandpoke/2010/12/the-free-model.html; imgur, "Facebook and You," *Imgur*, August 29, 2012, http://imgur.com/gallery/WiOMq.

72. Tim Lieder, "Oh Well. Fuck Amazon Vine," *LiveJournal*, August 19, 2013, http://marlowe1.livejournal.com/2129691.html.

73. Uri Gneezy and Aldo Rustichini, "A Fine Is a Price," *Journal of Legal Studies* 29, no. 1 (January 2000): 1–17, doi:10.1086/468061; Trevor Pinch and Filip Kesler, *How Aunt Ammy Gets Her Free Lunch: A Study of the Top-Thousand Customer Reviewers at Amazon.com*, Research Report, 2011, 45, http://www.freelunch.me/filecabinet/HowAuntAmmyGetsHerFreeLunch-FINAL.pdf.

74. E.Z., "An Interview with an Amazon Reviewer," June 2012.

Chapter 4: Improved

1. Richard I. Garber, "The Fourteen Worst Human Fears in the 1977 *Book of Lists*: Where Did This Data Really Come From?," *Joyful Public Speaking*, October 27, 2009, http://joyfulpublicspeaking.blogspot.com/2009/10/14-worst-human-fears-according-to-1977.html.

2. Kathleen Fitzpatrick, "If You Can't Say Anything Nice," *Planned Obsolescence*, January 25, 2013, http://www.plannedobsolescence.net/if-you-cant-say-anything-nice.

3. Matthew Kleinosky, "The Art of Effective Evaluation: Notes," *Gates to Excellence Toastmasters*, February 20, 2008, http://gatestoexcellence.org/Evals/ArtofEffectiveEvaluation_Notes.doc; Andrew Dlugan, "Delivering Effective

Speech Evaluations," *Six Minutes: Public Speaking and Presentation Skills Blog*, January 19, 2013, http://sixminutes.dlugan.com/speech-evaluation-2-art-of -delivering-evaluations.

4. Norbert Wiener, *Cybernetics: or Control and Communication in the Animal and the Machine*, 2nd ed. (Cambridge, MA: MIT Press, 1965), http://en.wikiversity .org/wiki/1948/Wiener; "feedback," *Oxford English Dictionary Online* (Oxford: Oxford University Press, 2012).

5. Michael Scriven, *The Methodology of Evaluation*, ed. Ralph Winfred Tyler, Robert Mills Gagné, and Michael Scriven (Chicago: Rand McNally, 1967); D. Royce Sadler, "Formative Assessment and the Design of Instructional Systems," *Instructional Science* 18 (1989): 120–122.

6. "Pedagogy," *Coursera*, February 13, 2013, https://www.coursera.org/about/ pedagogy.

7. Laura Gibbs, "About Me," *Pbworks*, February 19, 2013, http://onlinecourselady .pbworks.com/w/page/12783437/aboutme.

8. Audrey Watters, "The Problems with Peer Grading in Coursera," *Inside Higher Ed*, August 27, 2012, http://www.insidehighered.com/blogs/hack-higher -education/problems-peer-grading-coursera.

9. Laura Gibbs, "Done (More or Less)," *Coursera Fantasy*, September 3, 2012, http://courserafantasy.blogspot.com/2012/09/done-more-or-less.html.

10. Laura Gibbs, "Continuing Problems with Peer Feedback," *Coursera Fantasy*, August 24, 2012, http://courserafantasy.blogspot.com/2012/08/continuing -problems-with-peer-feedback.html.

11. Geoffrey J. Kennedy, "Peer-Assessment in Group Projects: Is It Worth It?," in *Proceedings of Australasian Computing Education Conference*, ed. Alison Young and Denise Tolhurst, vol. 42 (Australian Computer Society, 2006), http://crpit .com/confpapers/CRPITV42Kennedy.pdf; Christina M. Cestone, Ruth E. Levine, and Derek R. Lane, "Peer Assessment and Evaluation in Team-Based Learning," *New Directions for Teaching and Learning* 2008, no. 116 (2008): 69–78.

12. synecdochic, "'Cult of Nice' vs. 'Cult of Mean,' Round 2847, Fight," *Dreamwidth*, July 23, 2008, http://synecdochic.dreamwidth.org/235531.html.

13. Helen Nissenbaum, *Privacy in Context: Technology, Policy, and the Integrity of Social Life* (Stanford, CA: Stanford Law Books, 2009), 3; danah boyd, "Taken Out of Context: American Teen Sociality in Networked Publics" (Berkeley: University of California Berkeley, 2008), 34, http://www.danah.org/papers/ TakenOutOfContext.pdf.

14. psthebirdbites, "Sporking," *Urban Dictionary*, January 10, 2012, http://www .urbandictionary.com/define.php?term=Sporking&defid=6356771.

15. synecdochic, "'Cult of Nice' vs. 'Cult of Mean,' Round 2847, Fight."

16. Das Mervin, "Das_Sporking: Profile," *The Sporkings of Das Mervin and Company*, June 23, 2009, http://das-sporking.livejournal.com/profile.

17. Sarah Fay, "Book Reviews: A Tortured History," *The Atlantic*, April 25, 2012, http://www.theatlantic.com/entertainment/archive/2012/04/book-reviews

-a-tortured-history/256301; Zadie Smith, "This Is How It Feels to Me," *The Guardian*, October 13, 2001, http://www.theguardian.com/books/2001/oct/13/fiction.afghanistan.

18. David Denby, *Snark: It's Mean, It's Personal, and It's Ruining Our Conversation* (New York: Simon & Schuster, 2009), 2, 57.

19. Dragon Scholar, "Sporking, MSTing, and Mocking in Fandom," *Fanthropology: The Study of Fandom*, February 26, 2005, http://fanthropology.livejournal.com/18670.html.

20. Miriam Heddy, "When Writer and Beta Collide (and Why They Should)," *Fanfic Symposium*, February 11, 2013, http://www.trickster.org/symposium/symp62.html.

21. Robert A. Baron, "Negative Effects of Destructive Criticism: Impact on Conflict, Self-Efficacy, and Task Performance," *Journal of Applied Psychology* 73, no. 2 (1988): 199.

22. Stacey R. Finkelstein and Ayelet Fishbach, "Tell Me What I Did Wrong: Experts Seek and Respond to Negative Feedback," *Journal of Consumer Research* 39, no. 1 (2012): 35, doi:10.1086/661934.

23. Valerie J. Shute, *Focus on Formative Feedback*, Research Report (Princeton, NJ: Educational Testing Service, March 2007), http://www.ets.org/Media/Research/pdf/RR-07-11.pdf.

24. Clifford Nass and Corina Yen, *The Man Who Lied to His Laptop: What Machines Teach Us about Human Relationships* (New York: Penguin Group, 2010), 32, 55.

25. Arlie Russell Hochschild, *The Managed Heart: Commercialization of Human Feeling* (Berkeley: University of California, 2003), http://caringlabor.files.wordpress.com/2012/09/the-managed-heart-arlie-russell-hochschild.pdf.

26. Bill Henderson and André Bernard, *Pushcart's Complete Rotten Reviews and Rejections: A History of Insult, a Solace to Writers* (New York: Puschart Press, 1998), 12, 166.

27. Philip Ball, "Rejection Improves Eventual Impact of Manuscripts," *Nature News & Comment*, October 11, 2012, http://www.nature.com/news/rejection-improves-eventual-impact-of-manuscripts-1.11583.

28. Joyce Shor Johnson, "The Write Joyce: After the Beta Reader Reads! Eleven Things You Can Do," *Thewritejoyce*, June 1, 2011, http://thewritejoyce.blogspot.com/2011/06/after-beta-reader-reads-11-things-you.html.

29. Heather M. Whitney, "Reader Feedback: When Student Evaluations Are Just Plain Wrong," *ProfHacker*, May 17, 2011, http://chronicle.com/blogs/profhacker/reader-feedback-when-student-evaluations-are-just-plain-wrong/33378.

30. Chris McDonough, "In Praise of Complaining," *Plope*, January 1, 2012, http://plope.com/Members/chrism/in_praise_of_complaining.

31. Patricia Minicucci, "Critique Etiquette," *photo.net Site Help Forum > Photo Critique and Rating*, March 5, 2005, http://photo.net/site-help-forum/00BNwY.

32. Joseph Reagle, "Revenge Rating and Tweak Critique at Photo.net," in *Evaluating Creative Production in Digital Environments* (New York: Routledge, 2014), http://reagle.org/joseph/2013/photo/photo-net.html.

33. James Jerome Gibson, *The Ecological Approach to Visual Perception* (Boston, MA: Houghton Mifflin, 1979), http://books.google.com/books/about/The_Ecological_Approach_to_Visual_Percep.html?id=BJGCuje64FcC; Donald Norman, *The Design of Everyday Things* (New York: Doubleday, 1989).

34. "Wikipedia: Be Bold," *Wikipedia*, February 20, 2013, http://en.wikipedia.org/?oldid=539303557.

35. Greg Kroah-Hartman, "And People Wonder Why Kernel Maintainers Are Grumpy," *Google+*, April 2, 2012, https://plus.google.com/111049168280159033135/posts/5YtkxtuRXTy.

36. Linus Torvalds, "LINUX Is Obsolete," *Google Groups Comp.os.minix*, January 29, 1992, https://groups.google.com/forum/#!topic/comp.os.minix/wlhw16QWltI.

37. Linus Torvalds, "Load Keys from Signed PE Binaries," *LKML*, February 21, 2013, https://lkml.org/lkml/2013/2/21/228.

38. Evan Prodromou, "Linus Torvalds You Have a Terrible Temper and You Treat People Really Badly," *Google+*, March 1, 2013, https://plus.google.com/104323674441008487802/posts/Ew2ynALJd6R.

39. synecdochic, "'Cult of Nice' vs. 'Cult of Mean,' Round 2847, Fight."

40. Lunacy, "The Art of Critiquing," *Lunacy Reviews*, November 3, 2012, http://www.lunacyreviews.com/critique.php.

41. Sarah Bakewell, *How to Live: Or a Life of Montaigne in One Question and Twenty Attempts at an Answer*, reprint (New York: Other Press, 2011), 223; Michel Eyquem de Montaigne, *The Complete Works of Montaigne: Essays, Travel Journal, Letters*, trans. Donald M. Frame (Stanford, CA: Stanford University Press, 1958), 822, 642, 825.

Chapter 5: Alienated

1. "Fail," *Know Your Meme*, August 18, 2012, http://knowyourmeme.com/memes/fail.

2. Elizabeth Bear, "Whatever You're Doing, You're Probably Wrong," *Throw Another Bear in the Canoe*, January 12, 2009, http://matociquala.livejournal.com/1544111.html.

3. Avalon Willow, "Open Letter to Elizabeth Bear," *Seeking Avalon*, January 13, 2009, http://seeking-avalon.blogspot.com/2009/01/open-letter-to-elizabeth-bear.html.

4. stealthcomic, "BBS B.S." *Dave Ex Machina*, June 11, 2005, http://www.daveexmachina.com/wordpress/?p=996&cpage=1#comment-230.

5. Linus Torvalds, "LINUX Is Obsolete," *Google Groups Comp.os.minix*, January 29, 1992, https://groups.google.com/forum/#!topic/comp.os.minix/wlhw16QWltI.

6. Hongjie Wang, "Flaming: More Than a Necessary Evil for Academic Mailing Lists," *Electronic Journal of Communication* 6, no. 1 (1996), http://www.cios.org/EJCPUBLIC/006/1/00612.HTML.

7. Sara Kiesler, Jane Siegel, and Timothy W. McGuire, "Social Psychological Aspects of Computer-Mediated Communication," *American Psychologist* 39, no. 10 (1984): 1123–1134.

8. Philip G. Zimbardo, "The Human Choice: Individuation, Reason, and Order versus Deindividuation, Impulse, and Chaos," in *1969 Nebraska Symposium on Motivation*, 237–307, ed. Arnold and D. Levine (Lincoln: University of Nebraska, 1969); Edward Diener et al., "Effects of Deindividuation Variables on Stealing among Halloween Trick-or-Treaters," *Journal of Personality and Social Psychology* 33, no. 2 (1976): 178–183.

9. Tatsuya Nogami, "Reexamination of the Association between Anonymity and Self-Interested Unethical Behavior in Adults," *Psychological Record* 59 (2009): 259–272.

10. Random and Odd, "Internet Balls," *Urban Dictionary*, April 23, 2008, http://www.urbandictionary.com/define.php?term=internet%20balls; Penny Arcade, "Green Blackboards (and Other Anomalies)," *Penny-Arcade*, March 19, 2004, http://www.penny-arcade.com/comic/2004/03/19.

11. Cass R. Sunstein, "Boycott the Daily Me! Yes, the Net Is Empowering. But It Also Encourages Extremism—and That's Bad for Democracy," *Time Magazine*, June 4, 2001, http://content.time.com/time/magazine/article/0,9171,1957482,00.html; Eli Pariser, "The Filter Bubble," *Books* (Penguin Books, April 24, 2012); Ashley A. Anderson et al., "The 'Nasty Effect': Online Incivility and Risk Perceptions of Emerging Technologies," *Journal of Computer-Mediated Communication*, February 19, 2013, doi:10.1111/jcc4.12009.

12. Andrew, "The Troller's FAQ," *Internet Archive*, 2003, http://web.archive.org/web/20030105223101/http://www.altairiv.demon.co.uk/afaq/posts/trollfaq.html.

13. Susan Herring et al., "Searching for Safety Online: Managing 'Trolling' in a Feminist Forum," *Information Society* 18, no. 5 (October 2002): 377; Judith S. Donath, "Identity and Deception in the Virtual Community," in *Communities in Cyberspace*, ed. Peter Kollock and Mark Smith, 27–58 (London: Routledge, 1998), 14, http://smg.media.mit.edu/people/judith/Identity/IdentityDeception.html; Meatball, "AssumeGoodFaith," *Meatball Wiki*, September 16, 2006, http://meatballwiki.org/wiki/AssumeGoodFaith; mathew, "The Definitive Guide to Trolls," *Ubuntu Forums*, May 19, 2006, http://ubuntuforums.org/showthread.php?p=1032102.

14. E. Gabriella Coleman, "Phreaks, Hackers, and Trolls," in *The Social Media Reader*, ed. Michael Mandiberg, 109–110 (New York: New York University Press, 2011).

15. Stephen D. Reicher, Russell Spears, and Tom Postmes, "A Social Identity Model of Deindividuation Phenomena," *European Review of Social Psychology* 6 (1995); 161–198; Susan E. Watt, Martin Lea, and Russell Spears, "How Social Is Internet Communication? A Reappraisal of Bandwidth and Anonymity Effects," in *Virtual Society? Get Real! Technology, Cyberbole, Reality*, ed. Steve Woolgar, 61–77 (Oxford: Oxford University Press, 2003).

16. Robert Johnson and Leslie Downing, "Deindividuation and Valence of Cues: Effects on Prosocial and Antisocial Behavior," *Journal of Personality and Social Psychology* 37, no. 9 (1979): 1532–1538.

17. Philip Zimbardo, *The Lucifer Effect: Understanding How Good People Turn Evil* (New York: Random House, 2007).

18. Erin E. Buckels, Paul D. Trapnell, and Delroy L. Paulhus, "Trolls Just Want to Have Fun," *Personality and Individual Differences*, 2014, http://scottbarrykaufman .com/wp-content/uploads/2014/02/trolls-just-want-to-have-fun.pdf.

19. Ibid.

20. Whitney Phillips, "LOLing at Tragedy: Facebook Trolls, Memorial Pages and Resistance to Grief Online," *Firstmonday* 16, no. 12 (December 5, 2011), http:// firstmonday.org/htbin/cgiwrap/bin/ojs/index.php/fm/article/view/3168/3115.

21. Danielle Keats Citron, *Hate 3.0: A Civil Rights Movement for the Digital Age* (Cambridge, MA: Harvard University Press, 2013), 12.

22. Frank Paynter, "Mere Anarchy," *Listics Review*, April 8, 2007, http://web .archive.org/web/20070408234926/http://listics.com/20070326984.

23. Kathy Sierra, "A Very Sad Day," *Internet Archive*, March 26, 2007, http://web .archive.org/web/20070408002735/http://headrush.typepad.com/whathappened .html.

24. Chris Locke, "Re Kathy Sierra's Allegations," *Rageboy*, March 27, 2007, http://www.rageboy.com/2007/03/re-kathy-sierras-allegations.html.

25. Chris Locke and Kathy Sierra, "Kathy Sierra—Chris Locke—Coordinated Statements," *Rageboy*, April 2, 2007, http://rageboy.com/statements-sierra-locke .html.

26. weev, "Kathy Sierra," *Full Disclosure*, March 28, 2007, http://www.gossamer -threads.com/lists/fulldisc/full-disclosure/52577; Kathy Sierra, "State Machinery for State Machines," *TechCrunch*, June 25, 2013, http://techcrunch.com/2013/ 06/23/state-machinery-for-state-machines/?hubRefSrc=permalink#lf_comment =82508429.

27. Susan C. Herring, "Gender and Democracy in Computer-Mediated Communication," *Electronic Journal of Communication* 3, no. 2 (1993): 13, http:/ /www.cios.org/EJCPUBLIC/003/2/00328.HTML; Susan C. Herring, "Politeness in Computer Culture: Why Women Thank and Men Flame," in *Cultural Performances: Proceedings of the Third Berkeley Women and Language Group* 278– 294 (Berkeley, CA: Berkeley Women and Language Group, 1994), 291–292.

28. Locke and Sierra, "Kathy Sierra—Chris Locke—Coordinated Statements."

29. Andrea Weckerle, *Civility in the Digital Age: How Companies and People Can Triumph over Haters, Trolls, Bullies, and Other Jerks* (Indianapolis, IN: QUE, 2013), 100, http://my.safaribooksonline.com/book/communications/9780133134995.

30. Adam Lee, "Free Speech vs. Freeze Peach," *Patheos*, May 31, 2013, http://www.patheos.com/blogs/daylightatheism/2013/05/free-speech-vs-freeze-peach.

31. Joey, "The Stones of Time Is on My Side," *Misobservers*, April 2, 2007, http://misobserver.wordpress.com/2007/04/08/25.

32. Albert Bandura et al., "Mechanisms of Moral Disengagement in the Exercise of Moral Agency," *Journal of Personality and Social Psychology* 71, no. 2 (1996): 364–374.

33. mr-hank, "Hi, I'm the Guy Who Made a Comment about Big Dongles. First of All I'd Like to S …," *Hacker News*, March 19, 2013, https://news.ycombinator.com/item?id=5398681.

34. PyCon, "Code of Conduct," *PyCon US 2013*, April 1, 2013, https://us.pycon.org/2013/about/code-of-conduct.

35. Adria Richards, "Forking and Dongle Jokes Don't Belong at Tech Conferences," *But You're a Girl*, March 18, 2013, http://web.archive.org/web/20130508151159/http://butyoureagirl.com/14015/forking-and-dongle-jokes-dont-belong-at-tech-conferences.

36. mr-hank, "Hi, I'm the Guy Who Made a Comment about Big Dongles. First of All I'd Like to S…."

37. Gayle Laakmann McDowell, "Digging beneath the Surface: That Amanda Blum Article on Adria Richards Is Not What It Seems," *Technology Woman*, March 24, 2013, http://www.technologywoman.com/2013/03/24/digging-beneath-the-surface-that-amanda-blum-article-on-adria-richards-is-not-what-it-seems.

38. Emma A. Jane, "Your a Ugly, Whorish, Slut," *Feminist Media Studies*, 2012, 3, http://dx.doi.org/10.1080/14680777.2012.741073.

39. Troll Faq, "The Subtle Art of Trolling," *Urban75*, 1998, http://www.urban75.com/Mag/troll.html; Kevin Poulsen, "Hackers Assault Epilepsy Patients via Computer," *Wired*, March 28, 2008, http://www.wired.com/politics/security/news/2008/03/epilepsy.

40. Michael S. Bernstein et al., "4chan and /B/: An Analysis of Anonymity and Ephemerality in a Large Online Community," in *Proceedings of Proceedings of the Fifth International Conference on Weblogs and Social Media* (Association for the Advancement of Artificial Intelligence, 2011), http://www.aaai.org/ocs/index.php/ICWSM/ICWSM11/paper/download/2873/4398; Coleman, *Phreaks, Hackers, and Trolls*, 110–111.

41. Anonymous, "Richards Threats," *Pastebin*, March 21, 2013, http://pastebin.com/ubmznGhn.

42. Jim Franklin, "SendGrid Statement," *Sendgrid*, March 21, 2013, http://sendgrid.com/blog/sendgrid-statement; Jim Franklin, "A Difficult Situation," *Sendgrid*, March 21, 2013, http://sendgrid.com/blog/a-difficult-situation.

43. Stop the GR Bullies, "Why It's Time to Stop the Goodreads Bullies," *Huffington Post*, July 20, 2012, http://www.huffingtonpost.com/stop-the-gr-bullies/stop-goodreads-bullies_b_1689661.html.

44. Foz Meadows, "Goodreads and the Rise of Digital Reviewing," *Huffington Post*, July 25, 2012, http://www.huffingtonpost.com/foz-meadows/goodreads-and-the-rise-of_b_1700171.html.

45. "The List 2: Proof of Bullies on Goodreads," *Carroll Bryant*, August 25, 2012, http://carrollbryant.blogspot.com/2012/08/the-list-2-proof-of-bullies-on-goodreads.html.

46. Alice Marwick and danah boyd, "The Drama! Teen Conflict, Gossip, and Bullying in Networked Publics," paper presented at the A Decade in Internet Time: Symposium on the Dynamics of the Internet and Society, University of Oxford, September 2011, http://papers.ssrn.com/sol3/papers.cfm?abstract_id=1926349.

47. "How to Use the Author Program," *Goodreads*, May 6, 2013, http://www.goodreads.com/author/how_to.

48. Foz Meadows, "Bullying & Goodreads," *shattersnipe: malcontent & Rainbows*, July 10, 2012, http://fozmeadows.wordpress.com/2012/07/10/bullying-goodreads.

49. Bullies, "Why It's Time to Stop the Goodreads Bullies."

50. Andrew Losowsky, "Stop the GR Bullies: An Explanation," *Huffington Post*, July 20, 2012, http://www.huffingtonpost.com/andrew-losowsky/stop-the-gr-bullies-an-ap_b_1690134.html; Meadows, "Bullying & Goodreads."

51. Robin M. Kowlaski, Susan P. Limber, and Patricia W. Agatson, *Cyberbullying: Bullying In the Digital Age* (Malden, MA: Wiley-Blackwell, 2012).

52. "Reviews Are Not for Authors," *Readers Have Rights*, March 27, 2013, http://stopthegrbullies.net/reviews-are-not-for-authors.

53. Foz Meadows, "Stop the GR Bullies: A Response," *Huffington Post*, July 20, 2012, http://www.huffingtonpost.com/foz-meadows/stop-the-gr-bullies-a-response_b_1690469.html.

54. Lucy, "What It's Like to Be Stalked," *Goodreads*, July 21, 2012, http://www.goodreads.com/story/show/309224-what-it-s-like-to-be-stalked?page=8.

55. Rebecca Watson, "About Mythbusters, Robot Eyes, Feminism, and Jokes," *YouTube*, June 20, 2011, http://www.youtube.com/watch?v=uKHwduG1Frk; Rebecca Watson, "Sexism in the Skeptic Community: I Spoke Out, Then Came the Rape Threats," *Slate*, October 2012, http://www.slate.com/articles/double_x/doublex/2012/10/sexism_in_the_skeptic_community_i_spoke_out_then_came_the_rape_threats.html.

56. "RaceFail '09," *Fanlore*, February 22, 2013, http://fanlore.org/wiki/RaceFail_'09.

57. Richard Dawkins and Neil deGrasse Tyson, "Dawkins vs. Tyson," *YouTube*, November 22, 2006, http://www.youtube.com/watch?v=-_2xGIwQfik.

58. PZ Myers, "In Defense of the Commentariat," *Free Thought Blogs: Pharyngula*, December 27, 2012, http://freethoughtblogs.com/pharyngula/2012/12/27/in-defense-of-the-commentariat.

59. Nathan Hevenstone, "The Terrifying Pharyngula Commentariat," *Atheism, Music, and More*, December 28, 2012, http://natehevens.wordpress.com/2012/12/28/the-terrifying-pharyngula-commentariat.

60. Space Lemur Kirt, "#INeedMasculismBecause the … ," *Twitter*, February 8, 2013, https://twitter.com/xiombarg/status/299984316723785728.

61. Jason Thibeault, "Crashing the #FTBullies Hashtag," *Free Thought Blogs: Lousy Canuck*, July 4, 2012, http://freethoughtblogs.com/lousycanuck/2012/07/04/crashing-the-ftbullies-hashtag.

62. Elizabeth Bear, "Cease Fire," *Throw Another Bear in the Canoe*, March 5, 2009, http://matociquala.livejournal.com/1582583.html.

63. miss maggie, "sees Fire," *Bossy Marmalade*, March 4, 2009, http://bossymarmalade.dreamwidth.org/465929.html.

64. Rebecca Watson, "Why I Won't Be at TAM This Year," *Skepchick*, June 1, 2012, http://skepchick.org/2012/06/why-i-wont-be-at-tam-this-year.

65. Jen McCreight, "Goodbye for Now," *Free Thought Blogs: Blag Hag*, September 4, 2012, http://freethoughtblogs.com/blaghag/2012/09/goodbye-for-now.

66. Anita Sarkeesian, "Tropes vs. Women in Video Games," *Kickstarter*, June 7, 2012, http://www.kickstarter.com/projects/566429325/tropes-vs-women-in-video-games/posts/242547.

67. Sarah O'Meara, "Internet Trolls Up Their Harassment Game with 'Beat Up Anita Sarkeesian,'" *Huffington Post*, July 6, 2012, http://www.huffingtonpost.co.uk/2012/07/06/internet-trolls-online-beat-up-anita-sarkeesian-game_n_1653473.html.

68. Anita Sarkeesian, "Image Based Harassment and Visual Misogyny," *Feminist Frequency*, July 1, 2012, http://www.feministfrequency.com/2012/07/image-based-harassment-and-visual-misogyny.

69. feministfrequency, "Damsel in Distress: Part 1," *YouTube*, March 7, 2013, http://www.youtube.com/watch?v=X6p5AZp7r_Q.

Chapter 6: Shaped

1. Dan Savage and Terry Miller, "It Gets Better," *YouTube*, September 21, 2010, http://www.youtube.com/watch?v=7IcVyvg2Qlo.

2. Rachel Simmons, "What Formspring Means for Friendship and Female Aggression," *Jezebel*, September 2, 2011, http://jezebel.com/5836918/what-formspring-means-for-friendship-and-female-aggression; Amar Toor, "Ask.fm Responds to Cyberbullying Controversy with New Safety Measures," *The Verge*, August 19,

2013, http://www.theverge.com/2013/8/19/4635784/ask-fm-unveils-new-safety
-measures-amid-cyberbullying-suicides; danah boyd, "Harassment by Q&A:
Initial Thoughts on Formspring.me," *Zephoria*, April 26, 2010, http://www
.zephoria.org/thoughts/archives/2010/04/26/harassment-by-qa-initial-thoughts
-on-formspring-me.html.

3. xgothemo99xx, "It Gets Better, I Promise!," *YouTube*, May 4, 2011, http://
www.youtube.com/watch?v=-Pb1CaGMdWk.

4. danah boyd, *It's Complicated: The Social Lives of Networked Teens* (New
Haven, CT: Yale University Press, 2014), 141, http://www.danah.org/books/
ItsComplicated.pdf; Elizabeth Englander, "Digital Self-Harm: Frequence, Type,
Motivations, and Outcomes," *Bridgew*, Massachusetts Aggression Reduc-
tion Center (MARC), June 2012, https://webhost.bridgew.edu/marc/DIGITAL
%20SELF%20HARM%20report.pdf.

5. Eric Spitznagel, "Louis C.K.: "Starvation Can Be Character Building,"" *Vanity Fair*, March 2, 2009, http://www.vanityfair.com/online/daily/2009/03/louis-ck
-starvation-can-be-character-building.

6. "Safety Tips," *Formspring*, July 2, 2013, http://www.formspring.me/about/
safety.

7. "Stupid YouTube Comments," *Stupid YouTube Comments*, October 12, 2006,
http://stupid-youtube-comments.blogspot.com.

8. Erving Goffman, *Behavior in Public Places: Notes on the Social Organization
of Gatherings* (New York: Free Press, 1963).

9. Amy L. Gonzales and Jeffrey T. Hancock, "Mirror, Mirror on My Facebook
Wall: Effects of Exposure to Facebook on Self-Esteem," *Cyberpsychology, Behavior, and a Social Networking* 14, no. 1 (2011), doi:10.1089/cyber.2009.0411.

10. SaRAWRR, "Selfie," *Urban Dictionary*, October 12, 2012, http://www
.urbandictionary.com/define.php?term=selfie.

11. Triana Lavey, quoted in Claire Pedersen, "(Don't) Like My Photo: Social
Media Spurring Plastic Surgery," *ABC News*, July 16, 2012, http://abcnews
.go.com/Health/dont-photo-social-media-spurring-plastic-surgery/story?id
=16788142.

12. Carl Elliott, *Better Than Well: American Medicine Meets the American
Dream* (New York: Norton, 2003), 40.

13. William Peters and Charlie Cobb, "A Class Divided (Transcript)," *PBS: Frontline*, March 26, 1985, http://www.pbs.org/wgbh/pages/frontline/shows/divided/
etc/script.html.

14. Claude M. Steele and Joshua Aronson, "Stereotype Threat and the Intellectual
Test Performance of African Americans," *Journal of Personality and Social Psychology* 69, no. 5 (November 1995): 797–811, doi:10.1037/0022-3514.69.5.797.

15. Andrew Mecca, Neil J. Smelser, and John Vasconcellos, *The Social Importance of Self Esteem* (Berkeley: University of California Press, 1986), vii, http://
/publishing.cdlib.org/ucpressebooks/view?docId=ft6c6006v5;chunk.id=0;doc
.view=print.

16. Roy F. Baumeister et al., "Exploding the Self-Esteem Myth," *Scientific American*, December 20, 2004, 4, http://www.scientificamerican.com/article/exploding -the-self-esteem.

17. Carol S. Dweck, "The Perils and Promises of Praise," *Educational Leadership 65*, no. 2 (October 2007): 34–39; Claudia M. Mueller and Carol S. Dweck, "Praise for Intelligence Can Undermine Children's Motivation and Performance," *Journal of Personality and Social Psychology 75*, no. 1 (July 1998): 33–52.

18. Po Bronson and Ashley Merryman, *Nurtureshock: New Thinking about Children* (New York: Twelve, 2009); George Will, "Self-Esteem, Self-Destruction," *TownHall.com*, May 4, 2010, http://townhall.com/columnists/georgewill/2010/ 03/04/self-esteem,_self-destruction.

19. Amy Chua, "Why Chinese Mothers Are Superior," *Wall Street Journal*, January 8, 2011, http://online.wsj.com/article/SB10001424052748704111150457605 9713528698754.html.

20. Florrie Ng, Eva M. Pomerantz, and Shui-Fong Lam, "European American and Chinese Parents' Response to Children's Success and Failure: Implications for Children's Response," *Developmental Psychology 43*, no. 5 (2008): 1239–1255.

21. Richard L. Bednar and Scott R. Peterson, *Self-Esteem: Paradoxes and Innovations in Clinical Theory and Practice*, 2nd ed. (Washington, DC: American Psychological Association, 1995), 85, 96.

22. Zara Stone, "South Korean High Schoolers Get Plastic Surgery for Graduation," *The Atlantic*, June 27, 2013, http://www.theatlantic.com/international/ archive/2013/06/south-korean-high-schoolers-get-plastic-surgery-for-graduation/ 277255.

23. Robert H. Frank and Philip J. Cook, *The Winner-Take-All Society: How More and More Americans Compete for Ever Fewer and Bigger Prizes, Encouraging Economic Waste, Income Inequality, and an Impoverished Cultural Life* (New York: Free Press, 1995), 1, 15.

24. Hui-Tzu Grace Chou and Nicholas Edge, "'They Are Happier and Having Better Lives Than I Am': The Impact of Using Facebook on Perceptions of Others' Lives," *Cyberpsychology, Behavior, and Social Networking 15*, no. 2 (February 2012): 117–121, doi:10.1089/cyber.2011.0324.

25. Hanna Krasnova et al., "Envy on Facebook: A Hidden Threat to Users' Life Satisfaction?," paper presented at the Eleventh International Conference on Wirtschaftsinformatik (WI), Leipzig, Germany, 2013, 7, 12, http://warhol .wiwi.hu-berlin.de/~hkrasnova/Ongoing_Research_files/WI%202013%20Final %20Submission%20Krasnova.pdf.

26. Ethan Kross et al., "Facebook Use Predicts Declines in Subjective Well-Being in Young Adults," *PLOS One 8*, no. 8 (August 14, 2013), http://dx.doi.org/10 .1371/journal.pone.0069841.

27. Charles de Montesquieu, quoted in Alexander H. Jordan et al., "Misery Has More Company Than People Think: Understanding the Prevalence of Others' Negative Emotions," *Personality and Social Psychology Bulletin 37*, no. 1 (January 2011): 120, doi:10.1177/0146167210390822.

28. Ibid., 132.

29. Junghyun Kim and Jong-Eun Roselyn Lee, "The Facebook Paths to Happiness: Effects of the Number of Facebook Friends and Self-Presentation on Subjective Well-Being," *Cyberpsychology, Behavior, and Social Networking* 14, no. 6 (2011), doi:10.1089/cyber.2010.0374; Jong-Eun Roselyn Lee et al., "Who Wants to Be 'Friend-Rich'? Social Compensatory Friending on Facebook and the Moderating Role of Public Self-Consciousness," *Computers in Human Behavior* 28, no. 3 (May 2012): 1036–1043.

30. Spitznagel, "Louis C.K."; Louis C.K., "Louis C.K. Hates Twitter," *YouTube*, November 4, 2011, http://www.youtube.com/watch?v=xSSDeesUUsU.

31. Siva Vaidhyanathan, "Generational Myth: Not All Young People Are Tech-Savvy," *Chronicle Review* 55, no. 4 (September 19, 2008): B7.

32. Eszter Hargittai, "Whose Space? Differences among Users and Non-Users of Social Network Sites," *Journal of Computer-Mediated Communication* 13, no. 1 (2007), doi:10.1111/j.1083-6101.2007.00396.x.

33. Roy Pea et al., "Media Use, Face-to-Face Communication, Media Multitasking, and Social Well-Being among Eight- to Twelve Year-Old Girls," *Developmental Psychology* 48, no. 2 (2012): 328, 335, doi:10.1037/a0027030.

34. Sherry Turkle, *Alone Together: Why We Expect More from Technology and Less from Each Other* (New York: Basic Books, 2010), 161, 266–268.

35. Lawrence Lessig, *Republic, Lost: How Money Corrupts Congress—and a Plan to Stop It* (New York: Twelve, 2011).

36. Elizabeth Currid-Halkett, "Why Narcissism Defines Our Time," *Wall Street Journal*, December 8, 2010, http://blogs.wsj.com/speakeasy/2010/12/08/why-narcissism-defines-our-time; Jessica Stillman, "Is Facebook Turning Generation Y into a Bunch of Narcissists?," *CBS News*, December 14, 2010, http://www.cbsnews.com/8301-505125_162-38943743/is-facebook-turning-generation-y-into-a-bunch-of-narcissists.

37. James Napoli, "Facebook 'Like' Button Replaced with 'Love I Never Got from My Parents' Button," *Huffington Post*, June 26, 2012, http://www.huffingtonpost.com/james-napoli/facebook-like-button-repl_b_1626061.html.

38. Carolyn C. Morf and Frederick Rhodewalt, "Unraveling the Paradoxes of Narcissism: A Dynamic Self-Regulatory Processing Model," *Psychological Inquiry* 12, no. 4 (2001): 178, doi:10.1207/S15327965PLI1204_1.

39. Jean M. Twenge and W. Keith Campbell, *The Narcissism Epidemic: Living in the Age of Entitlement* (New York: Free Press, 2010), 2, 4; Jean M. Twenge, "The 'Debate' about Narcissism Increasing: More Twists Than a Crime Novel," *Psychology Today*, May 12, 2010, http://www.psychologytoday.com/blog/the-narcissism-epidemic/201005/the-debate-about-narcissism-increasing-more-twists-crime-novel.

40. Twenge and Campbell, *The Narcissism Epidemic*, 117, 85.

41. Laura E. Buffardi and W. Keith Campbell, "Narcissism and Social Networking Web Sites," *Personality and Social Psychology Bulletin* 34, no. 10 (October 2008): 1303–1314, doi:10.1177/0146167208320061.

42. Soraya Mehdizadeh, "Self-Presentation 2.0: Narcissism and Self-Esteem on Facebook," *Cyberpsychology, Behavior, and Social Networking* 13, no. 4 (2010), doi:10.1089/cyber.2009.0257.

43. Bill Ray, "Latest Subject for Peer Review? You," *The Register*, March 11, 2009, http://www.theregister.co.uk/2009/03/11/person_rating; Molly Wood, "Unvarnished: Person Reviews or Trollfest?," *CNET News*, March 31, 2010, http://news.cnet.com/8301-31322_3-20001507-256.html; "Job-Review Sites: Honestly Unvarnished," *The Economist*, December 8, 2012, http://www.economist.com/news/business/21567985-how-help-employees-spill-beans-and-make-money-it-honestly-unvarnished.

44. Evelyn Rusli, "Unvarnished Becomes Honestly.com, Raises $1.2 Million and Opens the Floodgates," *TechCrunch*, October 19, 2010, http://techcrunch.com/2010/10/19/unvarnished-honestly-kazanjy-funding.

45. "How the KarmaFile Peer Review System Works," *KarmaFile*, July 30, 2013, http://karmafile.com/how-it-works; "Popular Questions about KarmaFile," *KarmaFile*, July 30, 2013, http://karmafile.com/feedback.

46. George Ritzer, "The 'McDonaldization' of Society," *Journal of American Culture* 6, no. 1 (1983): 103.

47. "How It Works," *Klout*, July 31, 2013, http://klout.com/corp/how-it-works.

48. Ibid.

49. Ekaterina Walter, "Kred Rewards and Its Code of Conduct," *The Next Web*, June 29, 2012, http://thenextweb.com/insider/2012/06/29/kred-rewards-learning-from-the-mistakes-of-klout-perks.

50. Seth Stevenson, "What Your Klout Score Really Means," *Wired*, April 24, 2012, http://www.wired.com/epicenter/2012/04/ff_klout/all/1.

51. Somini Sengupta, "When Sites Drag the Unwitting across the Web," *New York Times*, November 13, 2011, http://www.nytimes.com/2011/11/14/technology/klouts-automatically-created-profiles-included-minors.html.

52. Rohn Jay Miller, "Delete Your Klout Profile Now!," *Social Media Today*, November 9, 2011, http://socialmediatoday.com/rohnjaymiller/385168/delete-your-klout-profile-now.

53. Alexandra Samuel, "A Social Sanity Manifesto for 2012," *Harvard Business Review*, December 12, 2011, http://blogs.hbr.org/2011/12/a-social-sanity-manifesto-for.

54. Calvin Lee, quoted in Stevenson, "What Your Klout Score Really Means."

55. Gilad Lotan, "(Fake) Friends with (Real) Benefits," *Medium*, June 5, 2014, https://medium.com/i-data/fake-friends-with-real-benefits-eec8c4693bd3.

56. Sean Lind, "How to Rate Girls: The Base Ten Scale Defined," *Real Men Drink Whiskey*, June 13, 2011, http://www.realmendrinkwhiskey.com/how-to-rate-girls; HB Authority, "HB," *Urban Dictionary*, December 6, 2006, http://www.urbandictionary.com/define.php?term=HB; Vince Lin, "Decimal Rating Scale," *PUA Lingo*, December 23, 2010, http://www.pualingo.com/pua-definitions/decimal-rating-scale.

57. Nikol Lohr, "The Dick List," *Disgruntled Housewife*, August 7, 2002, http:/ /web.archive.org/web/20020807202347/http://www.thedicklist.com/cgi-bin/ ultimatebb.cgi.

58. Dan Slater, "A Million First Dates," *The Atlantic*, January 2, 2013, http:// www.theatlantic.com/magazine/archive/2013/01/a-million-first-dates/309195/ ?single_page=true.

59. Stephanie Rosenbloom, "No Scrolling Required at New Dating Sites," *New York Times*, April 13, 2012, http://www.nytimes.com/2012/04/15/fashion/no -scrolling-required-at-new-dating-sites.html.

60. Jenna Wortham, "The Value of a Facebook Friend? About Thirty-seven Cents," *New York Times*, January 9, 2009, http://bits.blogs.nytimes.com/2009/01/ 09/are-facebook-friends-worth-their-weight-in-beef; Jenna Wortham, "'Whopper Sacrifice' De-Friended on Facebook," *New York Times*, January 15, 2009, http:/ /bits.blogs.nytimes.com/2009/01/15/whopper-sacrifice-de-friended-on-facebook.

61. Howard Rheingold, *Net Smart: How to Thrive Online* (Cambridge, MA: MIT Press, 2012), 36, 74.

Chapter 7: Bemused

1. Ken Fisher, "More Signal, Less Noise: Cleaning Up Our Comments," *Ars Technica*, April 9, 2010, http://arstechnica.com/staff/palatine/2010/04/more-signal -less-noise-cleaning-up-our-comments.ars.

2. Lev Muchnik, Sinan Aral, and Sean J. Taylor, "Social Influence Bias: A Randomized Experiment," *Science* 341, no. 6146 (August 9, 2013), doi:10.1126/science.1240466.

3. "Frequently Asked Questions," *Thunderclap*, May 19, 2014, https://www .thunderclap.it/faq.

4. "Amazon.com: Customer Reviews: Logitech Wireless Presenter R400," *Amazon*, December 15, 2009, http://www.amazon.com/Logitech-Wireless-Presenter -Laser-Pointer/product-reviews/B002GHBUTK.

5. Avengur, "Saved My Kids Life? Four out of Five Stars," *Imgur*, May 3, 2012, http://imgur.com/gallery/GXfga.

6. KayBe, "KayBe's Reviews," *Canadian Tire*, January 27, 2011, http://reviews .canadiantire.ca/9045/KayBe/profile.htm.

7. GoCU 407, 12parsecs and imgurian776, "Saved My Kids Life? Four out of Five Stars," *Imgur*, May 3, 2012, http://imgur.com/gallery/GXfga.

8. Trollpatrol, ibid.

9. Sadwer2012smk, spacedout83, and be_mindful, "Saved My Kids Life? Four out of Five Stars," *Reddit*, May 1, 2012, http://www.reddit.com/r/funny/ comments/t1yzc/saved_my_kids_life_4_out_of_5_stars.

10. Brian Regan, "Hospital," *YouTube*, August 9, 2009, http://www.youtube .com/watch?v=j1kuIwXaV5o.

11. Chris Pasero and Margo McCaffery, *Pain Assessment and Pharmacologic Management* (St. Louis, MO: Mosby, 2010).

12. Rob Johns, "Likert Items and Scales," in *SQB Methods Fact Sheet 1* (Question Bank, 2010), http://surveynet.ac.uk/sqb/datacollection/likertfactsheet.pdf.

13. Jerry W. Lee et al., "Cultural Differences in Responses to a Likert Scale," *Research in Nursing and Health* 25, no. 4 (2002): 295–306.

14. Jade Wang, "Local Businesses Get New Rating System," *Google and Your Business*, May 15, 2013, http://googleandyourbusiness.blogspot.com/2013/05/local-businesses-get-new-rating-system.html.

15. Geeky_username, spacedout83, and sadwer, "Saved My Kids Life? Four out of Five Stars," *Reddit*, May 1, 2012, http://www.reddit.com/r/funny/comments/t1yzc/saved_my_kids_life_4_out_of_5_stars/c4ixk22.

16. David Amann, "David Amann's Review of Diplomatic Immunity," *Amazon*, April 26, 2002, http://www.amazon.com/review/R3AFFYRPM82BSP.

17. Laura Miller, "The Dreaded Amazon Breast Curve," *Salon*, August 30, 2012, http://www.salon.com/2012/08/30/the_dreaded_amazon_breast_curve.

18. Images SI, "Uranium Ore: Amazon.com: Industrial and Scientific," *Amazon*, May 14, 2009, http://www.amazon.com/dp/B000796XXM/ref=azfs_379213722_UraniumOre_1.

19. J. Ingham, "J. INGHAM's Review of Uranium Ore," *Amazon*, May 20, 2009, http://www.amazon.com/review/R21S3QGYW3SCQC.

20. M. Carrier, "Strange Amazon Recommendations," *Amazon*, March 3, 2006, http://www.amazon.com/Strange-Amazon-Recommendations/lm/R15W3LLZV2GF0V.

21. Matthew Sidor, "Review of Denon AKDL1 Dedicated Link Cable," *Amazon*, June 23, 2008, http://www.amazon.com/review/R1P37H5L0NVOAN; R. Blais, "Review of Denon AKDL1 Dedicated Link Cable," *Amazon*, June 20, 2008, http://www.amazon.com/review/R1AYN2007LL1M3.

22. Spencer E. Ante, "Amazon: Turning Consumer Opinions into Gold," *Business Week*, October 15, 2009, http://www.businessweek.com/magazine/content/09_43/b4152047039565.htm; Daniel Emery, "Joke Review Boosts T-Shirt Sales," *BBC News*, May 21, 2009, http://news.bbc.co.uk/2/hi/technology/8061031.stm; Peter Applebome, "Think a T-Shirt Can't Change Your Life? A Skeptic Thinks Again," *New York Times*, May 24, 2009, http://www.nytimes.com/2009/05/25/nyregion/25towns.html; Brian Govern, "Review of The Mountain Youth Three Wolf Moon Shirt," *Amazon*, November 10, 2008, http://www.amazon.com/review/R2XKMDXZHQ26YX.

23. Andrew Adam Newman, "Playmobil's Checkpoint Strikes Some as Too Real," *New York Times*, February 15, 2009, http://www.nytimes.com/2009/02/16/business/media/16playmobil.html; Playmobile, "Playmobil Security Check Point," *Amazon*, 2005, http://www.amazon.com/PLAYMOBIL%C2%AE-36138-Playmobil-Security-Check/dp/B0002CYTL2.

24. "Funny Reviews," *Amazon*, August 16, 2013, http://www.amazon.com/gp/feature.html?docId=1001250201.

25. Antonio Reyes and Paolo Rosso, "Mining Subjective Knowledge from Customer Reviews: A Specific Case of Irony Detection," in *Proceedings of Workshop on Computational Approaches to Subjectivity and Sentiment Analysis (WASSA)* (Madison, WI: Association for Computational Linguistics, 2011), http://www.aclweb.org/anthology/W/W11/W11-17.pdf.

26. Elicia Dover, "Paid to Tweet? Rainn Wilson's Del Taco Promo Controversy," *ABC News*, October 27, 2011, http://abcnews.go.com/blogs/entertainment/2011/10/paid-to-tweet-rainn-wilsons-del-taco-promo-controversy.

27. CBS, "Rep. Weiner: I Did Not Send Twitter Crotch Pic," *CBS News*, May 29, 2011, http://www.cbsnews.com/stories/2011/05/29/politics/main20067242.shtml.

28. Daily Mail Reporter, "Anthony Weiner Admits Twitter Photo Could Have Been One of His That Was Taken out of Context," *Daily Mail Online*, June 2, 2011, http://www.dailymail.co.uk/news/article-1393503/Anthony-Weiner-admits-Twitter-photo-taken-context.html.

29. Steven Levy, "How Early Twitter Decisions Led to Weiner's Downfall," *CNN*, June 14, 2011, http://www.cnn.com/2011/TECH/social.media/06/14/twitter.decision.weiner.wired/index.html.

30. Tracie Egan Morrissey, "Racist Teens Forced to Answer for Tweets about the 'Nigger' President," *Jezebel*, November 9, 2012, http://jezebel.com/5958993/racist-teens-forced-to-answer-for-tweets-about-the-nigger-president.

31. Clive Coleman, "Robin Hood Airport Tweet Bomb Joke Man Wins Case," *BBC News*, July 27, 2012, http://www.bbc.co.uk/news/uk-england-19009344.

32. Shelton Green, "Texas Teen Charged with Making Terroristic Threat after Online Joke," *KVUE News*, June 25, 2013, http://www.khou.com/news/texas-news/Texas-teen-charged-with-making-terroristic-threat-after-online-joke-212931111.html.

33. Ibid.

34. Katie Zezima, "The Secret Service Wants Software That Detects Social Media Sarcasm. Yeah, Sure It Will Work," *Washington Post*, June 3, 2014, http://www.washingtonpost.com/blogs/the-fix/wp/2014/06/03/the-secret-service-wants-software-that-detects-social-media-sarcasm-yeah-sure-it-will-work.

35. John D. Sutter, "NRA Tweeter Was 'Unaware' of Colorado Shooting, Spokesman Says," *CNN*, July 20, 2012, http://www.cnn.com/2012/07/20/tech/social-media/nra-tweet-shooting.

36. Stuart Hall, "Encoding/decoding," in *Culture, Media, Language*, ed. Stuart Hall et al. (London: Routledge, 1992).

37. Louis C.K., "@danieltosh Your Show Makes ... ," *Twitter*, July 10, 2012, https://twitter.com/louisck/status/222849861224108035.

38. Daniel Tosh, "The Point I Was Making Before ... ," *Twitter*, July 10, 2012, https://twitter.com/danieltosh/status/222796636559130624.

39. Louis C.K. and Jon Stewart, "Interview," *The Daily Show*, July 16, 2012, http://www.thedailyshow.com/watch/mon-july-16-2012/louis-c-k-.

40. Daniel Tosh, "All the Out of Context Misquotes … ," *Twitter*, July 10, 2012, https://twitter.com/danieltosh/status/222796532653629441.

41. "Project Implicit," *Project Implicit*, August 28, 2013, https://implicit.harvard .edu/implicit/demo/background/faqs.html.

42. Can Durance, "I Got Twitter! I Figured It's about Time I Started Exploring … ," *Facebook*, March 20, 2013, https://www.facebook.com/IFeakingLoveScience/ posts/310235755771339?comment_id=1457812&offset=500&total_comments =1472; Amanda Holpuch, "Popular Science Blog Is Run by a Woman—to the Surprise of Some on Facebook," *Guardian*, March 20, 2013, http://www.guardian.co .uk/science/us-news-blog/2013/mar/20/i-love-science-woman-facbook.

43. John Knox IV, "I Was Pumped about the Hunger … ," *Twitter*, March 22, 2012, https://twitter.com/JohnnyKnoxIV/status/182972362583580673; Dodai Stewart, "Racist Hunger Games Fans Are Very Disappointed," *Jezebel*, March 26, 2012, http://jezebel.com/5896408/racist-hunger-games-fans-dont-care-how -much-money-the-movie-made.

44. J. K. Trotter, "How Twitter Schooled an NYU Professor about Fat-Shaming," *The Atlantic Wire*, June 3, 2013, http://www.theatlanticwire.com/national/ 2013/06/how-twitter-schooled-nyu-professor-about-fat-shaming/65833; Kathryn Cusma, "Visiting NYU Lecturer Makes Waves with Offensive Tweet about Fat People," *New York Post*, July 16, 2013, http://nypost.com/2013/06/04/visiting -nyu-lecturer-makes-waves-with-offensive-tweet-about-fat-people; Karen Wentworth, "Professor Geoffrey Miller Censured by UNM," University of New Mexico, August 6, 2013, http://news.unm.edu/newsmedia/advisories/professor -geoffrey-miller-censured-by-unm.

45. Cynthia Boris, "Chipotle's Fake Twitter Hack: Can Customers Take a Joke?," *Marketing Pilgrim*, July 29, 2013, http://www.marketingpilgrim.com/2013/07/ chipotles-fake-twitter-hack-can-customers-take-a-joke.html.

46. "Fuck You Yelper," *Fuck You Yelper*, July 31, 2012, http://fuckyouyelper .tumblr.com; "Real People, Real Reviews—Some of the Time," *Yelpers Who Suck*, April 10, 2013, http://yelperswhosuck.com.

47. Hillary Dixler, "Watch Amy's Baking Company Boldly Speak Out against Reddit, Yelpers, and Online 'Bullies'—Video Interlude," *Eater National*, May 21, 2013, http://eater.com/archives/2013/05/21/watch-amys-baking-company-boldly -speak-out-against-reddit-yelpers-and-online-bullies.php.

48. Raphael Brion, "Kitchen Nightmares Restaurant Freaks Out on Facebook," *Eater National*, May 14, 2013, http://eater.com/archives/2013/05/14/kitchen -nightmares-facebook-freakout.php; Lindsay William-Ross, "Owners of Amy's Baking Company, Restaurant Featured on 'Kitchen Nightmares,' Show How Not to Use the Internet," *LAist*, May 14, 2013, http://laist.com/2013/05/14/owners_of _amys_baking_company_az_re.php.

49. Nina Golgowski, "AP Twitter Hackers 'Break News' That White House Explosions Have Injured Obama," *Mail Online*, April 12, 2013, http://www

.dailymail.co.uk/news/article-2313652/AP-Twitter-hackers-break-news-White-House-explosions-injured-Obama.html.

50. Boris, "Chipotle's Fake Twitter Hack."

51. Kat Chow, "Haters Gonna Hate, as Shown on a Map," *NPR*, June 1, 2013, http://www.npr.org/blogs/codeswitch/2013/05/30/187280870/haters-gonna-hate-as-shown-on-a-map; Monica Stephen, "Hate Map," Humboldt State University, 2013, http://users.humboldt.edu/mstephens/hate/hate_map.html.

52. Meghan Neal, "Twitter Knows Which Restaurants Are Getting People Sick," *Motherboard*, August 9, 2013, http://motherboard.vice.com/blog/twitter-knows-which-restaurants-are-getting-people-sick.

53. Anonymous, "When I Watch People I Get a Tingly Sensation in My Head?!?," *Is It Normal?*, April 12, 2009, http://isitnormal.com/story/when-i-watch-people-i-get-a-tingly-sensation-in-my-head-28079.

54. "Welcome," *AIHO*, March 13, 2010, http://web.archive.org/web/20100313153853/http://aiho.org.

55. Andrew, "The Story So Far: A Timeline with History," *The Unnamed Feeling*, September 6, 2011, http://theunnam3df33ling.blogspot.com/2011/09/asmr-story-so-far-timeline-with-history.html; "Autonomous Sensory Meridian Response," *Wikipedia*, September 2, 2013, http://en.wikipedia.org/?oldid=571212729; Don, "Edit #126,714 (Autonomous Sensory Meridian Response (ASMR))," *Know Your Meme*, July 6, 2012, http://knowyourmeme.com/memes/autonomous-sensory-meridian-response-asmr/edits/126714.

56. Andrea Seigel, "Transcript," *This American Life*, March 29, 2013, http://www.thisamericanlife.org/radio-archives/episode/491/transcript.

Chapter 8: Conclusion

1. Amanda Brennan, "Interview about Comments," September 2013.

2. Brian Carnell, "To Offer Discussion Groups or Not," *Blog*, September 18, 2003, http://brian.carnell.com/articles/2003/to-offer-discussion-groups-or-not; Mark Frauenfelder, "Welcome to the New Boing Boing!," *Boing Boing*, August 28, 2007, http://boingboing.net/2007/08/28/welcome-to-the-new-b.html; Xeni Jardin, "Online Communities Rot without Daily Tending by Human Hands," *Edge*, 2008, http://www.edge.org/q2008/q08_7.html#jardin.

3. Justin Cheng, Cristian Danescu-Niculescu-Mizil, and Jure Leskovec, "How Community Feedback Shapes User Behavior," *Arxiv*, May 6, 2014, 9, http://arxiv.org/abs/1405.1429.

4. Rob Beschizza, "Time to Turn Off Comments for Good." *Muckrack*, February 28, 2013, http://muckrack.com/Beschizza/statuses/307172805252894722; Rob Beschizza, "Can We Talk?," *Boing Boing*, June 27, 2013, http://boingboing.net/2013/06/27/can-we-talk.html.

5. Nathan Hornby, "Hitler Finds Out Boing Boing Is Killing Comments, Moving to Discourse," *Boing Boing*, June 30, 2013, http://boingboing.net/2013/06/27/hitler-finds-out-boing-boing-i.html#comment-946944196.

6. Nick Bilton, "Disruptions: Gawker Wants to Encourage More Voices Online, but with Less Yelling," *New York Times Bits*, September 22, 2013, http://bits .blogs.nytimes.com/2013/09/22/disruptions-gawker-wants-to-encourage-more -voices-online-but-with-less-yelling.

7. Jeff Atwood, "Coding Horror: Civilized Discourse Construction Kit," *Coding-horror*, February 5, 2013, http://www.codinghorror.com/blog/2013/02/civilized -discourse-construction-kit.html.

8. Jeff Atwood, "Please Read the Comments," *Coding Horror*, March 19, 2014, http://blog.codinghorror.com/please-read-the-comments.

9. Ta-Nehisi Coates, quoted in Nate Matias, "The Beauty and Terror of Commenting Communities: Ta-Nehisi Coates at the Media Lab," MIT Center for Civic Media, October 22, 2012, http://civic.mit.edu/blog/natematias/the-beauty-and -terror-of-commenting-communities-ta-nehisi-coates-at-the-media-lab.

10. Judith Donath, *The Social Machine: Designs for Living Online* (Cambridge, MA: MIT Press, 2014).

11. Casey Johnston, "YouTube Hilariously Impotent against ASCII Comment Pornographers," *Ars Technica*, November 26, 2013, http://arstechnica.com/ business/2013/11/youtube-tries-to-stem-the-flow-of-a-new-kind-of-terrible -comments; Sarah Perez, "YouTube Addresses Massive Spam Problem Following Rollout of Much-Maligned Google+ Commenting System," *TechCrunch*, November 6, 2013, http://techcrunch.com/2013/11/26/youtube-addresses-massive-spam -problem-following-rollout-of-much-maligned-google-commenting-system; Alex Hern, "YouTube Co-Founder Hurls Abuse at Google over New YouTube Comments," *The Guardian*, November 8, 2013, http://www.theguardian.com/ technology/2013/nov/08/youtube-cofounder-why-the-fuck-do-i-need-a-google -account-to-comment.

12. Saylor and deviousfires commenting in Nundu Janakiram and Yonatan Zunger, "Turning Comments into Conversations That Matter to You," *YouTube Creator Blog*, November 6, 2013, http://youtubecreator.blogspot.com/2013/11/ turning-comments-into-conversations.html.

13. YTCreators, "An Update on YouTube Comments," *YouTube Creator Blog*, November 25, 2013, http://youtubecreator.blogspot.com/2013/11/an-update-on -youtube-comments.html.

14. insert clever name here, "The Mother of All Self Links?," *MetaFilter*, May 22, 2014, http://www.metafilter.com/139305/The-Mother-of-All-Self-Links.

15. Xeni Jardin, quoted in Nadja Popovich, "Xeni Jardin: Standing Up to Breast Cancer in Public and in Private," *The Guardian*, October 15, 2013, http://www .theguardian.com/society/2013/oct/15/xeni-jardin-breast-cancer-public-private.

16. Mark Frauenfelder, "Interview about Comments," October 18, 2013.

17. Dave Winer, "No Comment," *Scripting News*, February 19, 2012, http:// scripting.com/stories/2012/02/19/noComment.html.

18. Trent Reznor, quoted in David Marchese, "Trent Reznor's Upward Spiral," *Spin*, September 3, 2013, http://www.spin.com/featured/trent-reznor-upward -spiral-nine-inch-nails-spin-cover-september-2013.

19. Raymond Williams, "Drama in a Dramatized Society" (1974), in *Raymond Williams on Television: Selected Writings*, ed. Alan O'Connor (Toronto: Routledge, 1989), 3–5.

20. Alexander Rose, "Invisible Glove," *Cool Tools*, September 5, 2013, http://kk.org/cooltools/archives/12806.

21. Cap Watkins, "Formspring: A Postmortem," *Blog*, March 28, 2013, http://blog.capwatkins.com/formspring-a-postmortem.

22. Formspring executive, quoted in Casey Newton, "Killer App: Why Do Anonymous Q&A Networks Keep Leading to Suicides?," *The Verge*, September 17, 2013, http://www.theverge.com/2013/9/17/4740902/no-good-answers-why-didnt-ask-fm-learn-from-the-formspring-suicides.

23. Jamie Doward, "Twitter under Fire after Bank Note Campaigner Is Target of Rape Threats," *The Observer*, July 27, 2013, http://www.theguardian.com/uk-news/2013/jul/27/twitter-trolls-threats-bank-notes-austen; Katrin Bennhold, "Bid to Honor Austen Is Not Universally Acknowledged," *New York Times*, August 4, 2013, http://www.nytimes.com/2013/08/05/world/europe/bid-to-honor-austen-is-not-universally-acknowledged.html; Caroline Criado-Perez, "Have the Police Failed to Record the Twitter Threats against Me?," *New Statesman*, September 5, 2013, http://www.newstatesman.com/voices/2013/09/have-police-failed-record-twitter-threats-against-me; Alexandra Topping, "Caroline Criado-Perez Deletes Twitter Account after New Rape Threats," *The Guardian*, September 6, 2013, http://www.theguardian.com/technology/2013/sep/06/caroline-craido-perez-deletes-twitter-account.

24. "Anita's Irony," *Wikia*, March 2, 2013, http://geekfeminism.wikia.com/wiki/Anita's_Irony.

25. Matt Smith, "'Grotesque' Misconduct Gets Danziger Bridge Verdicts Tossed," *CNN*, September 18, 2013, http://www.cnn.com/2013/09/17/justice/louisiana-danziger-bridge-case/index.html.

26. Bland v. Roberts, 857 F. Supp. 2d 599 (E.D. Va. 2012), http://www.dmlp.org/sites/citmedialaw.org/files/12-29-2011-Reply%20to%20Response%20Motion%20for%20Summary%20Judgment.pdf; Venkat Balasubramani and Eric Goldman, "Facebook 'Likes' Aren't Speech Protected by the First Amendment, Rules Judge," *Ars Technica*, April 28, 2012, http://arstechnica.com/tech-policy/news/2012/04/facebook-likes-arent-speech-protected-by-the-first-amendment-bland-v-roberts.ars; Bland v. Roberts, No. 12-1671, 2013 WL 5228033 (4th Cir. Sept. 18, 2013), 39–40, http://legaltimes.typepad.com/files/usca4-facebook.pdf.

27. "A. G. Schneiderman Announces Agreement with Nineteen Companies to Stop Writing Fake Online Reviews and Pay More Than $350,000 in Fines," Eric T. Schneiderman, September 23, 2013, http://www.ag.ny.gov/press-release/ag-schneiderman-announces-agreement-19-companies-stop-writing-fake-online-reviews-and.

28. Robert H. Frank and Philip J. Cook, *The Winner-Take-All Society: How More and More Americans Compete for Ever Fewer and Bigger Prizes, Encouraging Economic Waste, Income Inequality, and an Impoverished Cultural Life*

(New York: Free Press, 1995); George Ritzer, "The 'McDonaldization' of Society," *Journal of American Culture* 6, no. 1 (1983): 100–107.

29. Katherine Losse, *The Boy Kings: A Journey into the Heart of the Social Network* (New York: Free Press, 2012), 128.

30. Chris Welch, "Google's New Gesture Patents Could Let Glass Users 'Heart' Things in Real Life," *The Verge*, October 15, 2013, http://www.theverge.com/2013/10/15/4841764/google-gesture-patents-could-let-glass-users-heart-things-in-real-life.

31. Howard Rheingold, *Net Smart: How to Thrive Online* (Cambridge, MA: MIT Press, 2012), 36, 74.

Index